A Spin Around
The Universe

or

Natural Dialectic,

A Theory of Complementary Opposites:

And an Undergraduate Course With Test Qs and As.

Published by Merops Press

website: www.cosmicconnections.co.uk

also www.scienceandphilosphy.co.uk

www.scienceandthesoul.co.uk

Acknowledgements: Suzanne, Marianne, Emmanuel and Françoise.

Contents

Preface

Introducing a simple structure to reunify science and philosophy or, in other words, a way in which to generate an internally self-consistent philosophy regarding information, physics, psychology, biology and community.

While not a travelogue that ranges through the accumulated libraries, schools and experiments of human endeavour, this book cuts with ease to the heart of what they are after - universal pattern. Then, if you want elaboration from pattern to details, it indicates a way to logically look.

A Spin Around the Universe? Who doesn't like an easy ride? Five minutes to admire a painting, ten on an arousing piece of music or several hours before the magic of a movie screen. Passive absorption. Passing sensations. But is it possible to understand science and philosophy with quite such ease? Harder going needs a helping hand. Some students say that, first time through, an easy draft is needed.

So, to understand the roots of life and its stage, the universe, this book takes it easy - but with strings attached. These strings are end-notes called connections that radiate to link with elaborations in other books in the Science and Philosophy series. They will allow direct access to check out deeper questions that the text provokes; they will reveal the coherent trail of logic behind some apparent conflicts between science and philosophy; and challenge with some unexpected assertions. In short, this submission delivers *The Reunification of Science and Philosophy* with 170 stimulating questions and answers in a way easily convertible to a 2-term course of lectures and, if wished, seminars using (PowerPoint) slides.

To follow any trail you need instruments of navigation. To this end what follows introduces the format and implications of a fresh and reasonable compass - a simple structure called **Natural Dialectic**.

Natural is commonly construed as a characteristic of any material item not devised by the mind or produced by the hand of man.

Dialectic (or dialectical method) is to-fro discourse between opposing views in order to establish reasonable truth. You might start poles apart but, at length, agree to only partly disagree!

Their combination, Natural Dialectic is, effectively, a Theory of Complementary Opposites. It is one of different models seekers use to try and understand the universe into which each one of us is, without asking, born. Variously expressed at different times and places, the Dialectic's 'philosophical machine' reflects a binary, oscillatory framework within which nature seems to operate.

Any philosophical infrastructure, whether mathematical or verbal, is built of symbols, that is, of code or language. Language, involving a specific assignment and coherent arrangement of symbols, is the way that meaning is organised and information conveyed. In this case, could the binary structure of **Natural Dialectical Philosophy** accurately describe the *modus operandi* of cosmos? Could its spine of complementary opposites reflect the way our universe is built?

This is guidance by simplicity. But some of us don't take to 'grammar' very easily, so the structure of this Theory of Opposites, Natural Dialectic, has been transferred and now is waiting in Appendix 1.

Meanwhile, for those who plough straight on, there is also a Glossary and, of course, the informative trove of Google. In our age of information interest can always educate itself. For easy skimming, the devices of **bold**, underlines, *italics* and red script simply draw attention to salient points. *And don't forget to use the Glossary.* This also contains, for interest's sake, an explanation of some oriental Sanskrit terms.

You can also, if you wish, exploit extensive Connections/ Endnotes (indicated on the Contents page) by reaching back to companion volumes mentioned there in order to elaborate on queries that will almost inevitably be raised by this book's considerable abbreviation (see also *www.cosmicconnections.co.uk*). **Also, importantly, you can check your understanding against the *questions* at the end of every chapter with *answers* at the book's end.**

Through to its end the book is, as regards its metaphysic, scrupulously, religiously non-religious. It can well be read in conjunction with the One World Lecture Slide Set. This set of 4 volumes, with over 500 slide pictures, systematically elaborates each section of A Spin Around the Universe.

Finally, after careful and perhaps exciting inspection of these chapters (and possibly their correlated slide set) you may better judge whether Natural Dialectic's grammar well interprets nature's text and thereby accurately reflects the logic of creation. So jump aboard and ride this streamlined 'thought machine'. Intellectual seat-belt fastened, we shall travel far and fast from here…

Chapter 1: Introduction

A Spin Around The Universe? **This book's orbit round creation is, in practice, a year-long undergraduate course including 170 test and exam questions, a glossary and much material to stimulate seminars.** It is correlated with a lecture series and is, in this respect, a primer to help you more clearly, easily and succinctly understand what's going on. **The aim, in a nutshell, is to reunify science and philosophy**. This venture elicits **two subsidiary aims**. The first involves,

> **in the wisdom of Albert Einstein's words,**
>
> *"... seeking the simplest possible scheme of thought that will bind together the observed facts".*

Yes, indeed. And, secondly,

> *to generate a philosophical routine, a 'dynamic' within which physic and metaphysic are accommodated and may be reconciled.*

We underline it. At root, the idea is to introduce a simple structure within which to generate an internally self-consistent philosophy regarding information, physics, psychology, biology and community.

The second, correlated aim is to compare, as if through different angles of a prism, the logic of our material world (as explored by scientific study) with the perspective that snaps into view if a single immaterial element is added.

> *material   immaterial*

What is this metaphysical addition, one very well-known and, indeed, forming the *basis* of our current age? As the industrial age was based on the energy of steam, electricity and, finally, nuclear power so ours is based on information. *Information* **is the immaterial element, that is, the metaphysical addition.**

Soon we'll methodically compare two perspectives[1] - call them **materialism and holism** - and the logic of interpretations that derive from them *using three simple models* of a balance, that is, a scale along with concentric rings and a stepped pyramid called Mount Universe.

Models are non-verbal descriptions. They are pictures to hang concepts on. To work successfully their underlying 'grammar', that is, mode of operation must be clear. We shall spend time this first lesson exploring the conceptual vehicle or 'cosmic language' which they help illustrate. **If we**

want a fresh perspective we will need to understand this language and its mode of expression.

If correct, such language should generate an abstract, metaphysical machine, tight-knit, well-riveted by bolt and counter-bolt, the simplest working model of the universe. But what is such a vehicle called? How does this philosophical construction work? Its perspective is not mathematical; equations of a physicist may well describe the music of a song but numbers miss the whole point. Science has derived an objective, mathematical and exclusively *materialistic* interpretation of events; but how are facts poured into a holistic 'dynamic' so that we add the *immaterialistic* part of our comparison and make whole again?

At this point I should straightaway scotch the notion that any considerations concerning mind, the origins of the universe or life that are not entirely materialistic are somehow **'anti-science'**. If I offer a holistic interpretation of this wall, door or the air am I 'anti-scientific'? **Am I 'anti' non-conscious matter or energy - the preserve of scientific study?** Inclusive holism deals with both material *and* immaterial elements but to say it is 'anti' one of them is absurd.

Finally, let's recall that *all* of science and philosophy is concerned to address the most natural questions of a being that finds him or herself as a body on earth, a planet in space, for a brief spell of time.

Such enquiry needs answers reasonably argued within one of two basic frames of reference - either **materialism** with its sub-creeds of humanism, scientific atheism and so on; or **holism** (see Glossary), as old as recorded human thought, including metaphysic *and* a science of material reality, with its sub-creeds and various forms of faith as well. Since nature is pre-religious neither science nor Natural Dialectic, a holistic Theory of Complementary Opposites, is religious. Within a neutral, natural framework we'll ask:

> **What's it all about?**
> **Who exactly am I?**
> **Where did I come from?**
> **What am I doing here?**
> **Indeed, is there any purpose in life other than the satisfaction of animal needs?**
> **Then, where am I going?**
> **What happens, when the pot cracks, after death?**
> **And finally, of course, the questions of morality, law, society and government both political and personal.**

Now let's take a look at the **Primary Assumptions**[2] of materialism and holism. Natural Dialectic systematically compares the implications of each mind-set. Which assumption, aimed towards full truth and understanding, does the evidence best brace? What are the basic axioms and corollaries of the pair?

The Primary Axiom of Materialism **is that every object and event, including an origin of the universe and the nature of mind, are material alone; a few oblivious kinds of particles and forces compose all things.**

Although the universe appears to work by rules and to have been established in a very particular way, this appearance of order is in fact unplanned. Its invisible framework of regulation must have occurred by chance and, since inception, individual objects and events (called actualities) occur by chance as well.

Materialism's Primary Axiom **is that cosmos issued out of nothing; therefore, beyond this realm of physics there was only void; and life is an inconsequent coincidence, electric flickers of illusion in a lifeless, dark eternity.**

Although the universe appears to work by rules and to have been established in a very particular way, this appearance of order is in fact unplanned. Materialism's cosmic reason is, thereby, its own antithesis - unreason. Rules by chance, events by reflex. Oblivious matter is an aimless actor whence, by accident, all life rose. Life has, hasn't it, to be the offspring of non-conscious particles and forces?

Such axiom must apply to life. In this respect the *Primary Corollary of Materialism* **states, by the neo-Darwinian theory of evolution, that life forms are the product of the chemical abiogenesis of a first cell; and following that, by common descent, of a random generator (mutation) acted on by a filter called natural selection. Such evolution is an absolutely mindless, purposeless process. These notions are, from a materialistic perspective, fact so that this** *PCM* **is a fundamental** *mantra* **of materialism.**

Chance is the creator of diversity. Its scientific *aide-de-camp* is probability. **No matter what the odds against, the universe and life *must have* appeared by chance.** Order came about by, basically, chance. Without reason. No telling how exactly, just vague imprecation. Nothing is, perchance, impossible; the sole impossibility is that such a story of creation is impossible. This implacably materialistic, possibly nonsensical, narrative is rehashed in almost every modern textbook, journal and media broadcast.

A caveat. To materialistically presume that what is not material is not natural and, therefore, does not exist is a first order, pseudoscientific error. On the other hand, to 'pretend' metaphysic *does* exist is prejudiciously judged, by that same materialistic rationale, 'pseudoscience'. If, however, the basic nature of information is immaterial/ metaphysical then isn't to construe every IT program, engineering blueprint, artistic design, theory and even your own thoughts as an irrational, 'pseudoscientific' exercise? In which case, could our scientific outlook be lop-sided? Isn't balance needed? Is a fresh perspective possible?

> *Holism's axiom* is that realistic comprehension of the world includes *two* primary components - immaterial and material or, as obvious to everyone, mind and matter.
>
> A scientific world-view that does not profoundly and completely come to terms with the nature of conscious mind can have no serious pretension of wholeness.

Holism's Primary Axiom is, on the other hand, that realistic comprehension of the world includes *two* primary components - immaterial and material or, as commonly perceived, mind and matter.

Holism, therefore, simply adds immaterial, as a second fundamental ingredient, to material. Or, conversely, it adds material to immaterial. Since immaterial is not material it adds nothing physical at all. **But hence follows, it is argued, holism's powerful and impregnable validity.**

> The *Primary Corollary of Holism* states that the origin of irreducible, biological complexity is not an accumulation of 'lucky' accidents constrained by natural law and death.
>
> Forms of life are conceptual; they are, like any creation of mind, the product of purpose. Such assertion is, in the face of materialism, absolute anathema. *Yet, if materialism's first axiom is incomplete then every step that follows will lead further from original truth. An axiom that discounts the force of immaterial information may well be largely incomplete.*

Holistic logic must also apply to bio-logical life. **In this respect its Primary Corollary states that the origin of irreducible, codified (or highly informed) biological complexity is not an accumulation of 'lucky'**

accidents constrained by natural law and death. Forms of life are programmed, purposive and wholly dependent on an inherently immaterial element - information.

Such assertion is, in the face of materialism, absolute anathema. Yet, if materialism's first axiom is actually incomplete then every step that follows will lead further from full truth. *An axiom that discounts the force of information may well be largely incomplete.*

Furthermore, let us at the outset be completely clear - the basic assertions of both materialism and holism are philosophical; *neither* **party is a scientific one because neither can scientifically prove nor disprove the other.** To engage *either* stance therefore involves a leap of faith. Holism includes metaphysic; materialism of necessity excludes it. *In this case, if the holistic axiom that mind and matter are two different characters is true then holistic logic in its entirety is unassailable.*

Three main points arise.

1. Holistic axiom exacts a toll. We need answers about the following:[3]

(i)   **the nature of consciousness, sub-consciousness and non-consciousness.**

(ii)   **whether mind, either individual or universal, can exist independent of its body and, if so, the nature of its entry, attachment, influence, exit and disembodied condition.**

(iii)   **the interactive relationship of individual mind with body; the nature of any** *PSI* **(Psycho-Somatic Interface or, perhaps, quantum linkage) between mind and matter.**

(iv)   **the mechanism by which universal mind, if such metaphysical unifying factor exists, might inform non-conscious forces, particles and gross phenomena; the origin of physical constants, that is, the source of natural law.**

(v)   **the nature of physical and biological prototypes, homologies or, if any, archetypes.**

(vi)   **the question whether biology is informed by chance and aimless natural law or by design in accordance with such law; a wholesale reappraisal of the neo-Darwinian theory of evolution.**

2. We need to nail down the language (one not exclusively materialistic) in which holistic answers and arguments can be included - in this case holistic **Natural Dialectic.**

Some find learning a new language onerous. Grammar, let alone a philosophical type of grammar, lacks appeal. **Why, therefore, bother?**

We bother because the structure of Natural Dialectic is a simple, orderly and efficient vehicle in which to marshal facts, in terms of opposites, into a digital framework. This framework, being verbal (in the line of Lao Tse, the I Ching's Pa Kua and others) rather than mathematical (in the line of Liebniz, Boole, Braille or ASCII code) is imprecise compared to scientific formulae. Instead, it builds relationships, connections and equivalences in a way that is logical and self-consistent. These can sometimes trigger fresh realisations and perspectives, that is, *eureka* moments.

Why does this matter? *Firstly,* because it includes both physical *and,* inaccessible to mathematics, metaphysical components. The latter, including subjective experience, purpose and so forth, are central to us. They are our life. Nor can any scientific or other explanation that fails to explain consciousness be complete.

Secondly, it can therefore be used to systematically compare the implications of materialism as opposed to supra-materialism (holism).

Thirdly, its binary, oscillatory logic may well turn out to reflect the way nature works. If so its columns, called stacks, are the backbone of this Theory of ComplementaryOpposites. Its polarities represent not only human but the cosmic spine; or, if you like, the system generates a robust body of philosophy that accurately reflects the order of the cosmos. It generates an abstract, metaphysical machine, tight-knit, well riveted by bolt and counter-bolt, the simplest working model of the universe.

It may, therefore, pay dividends to understand the very simple way it works. Let's try. Let's start with three pictorial models.

3. **Using these three universal models of creation:**[4]

   a) You can think of cosmos as a **pair of scales**.

high concentration/
active extreme

Source/ Origin;
Prior Super-state;
Pre-active Balance; Full Potential

discharge; diffusion;
entropy of energy or
information

informative or
energetic stimulus/
action

Pivot/ Zero-point

exhaustion; oblivion;
flat impotence; no more possibility;
substate condition; sink

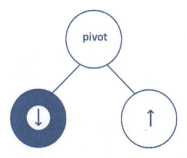

Now inspect this triplex 'stack':

↓ descent	Pivot	ascent ↑
swing down	Centre-point	swing up
gravity	Equilibrium	levity

Does this triplex, formed by pivot and two linked arms represented as vectors (up (↑) and (↓) down), constitute a trinary or binary system? In universal terms, the Motionless Centre-point represents Essence and the two antagonistic yet complementary vectors represent ever-changing existence.

$$↓ \ existence \ ↑ \qquad Essence$$

If this Essential Axis is itself antagonistic yet complementary to peripheral existence then this pair compose the primary pair of opposites; and, although non-vectored Essence is singular, existence is dual, split by its vectors.

b) You can think of cosmos in *dynamic* terms of <u>concentric rings or, three-dimensionally, as spheres.</u>

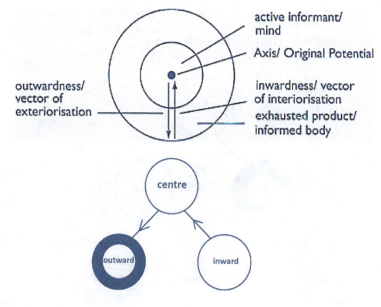

13

Diffusion and concentration. This figure of concentric rings includes antagonistic vectors from and to its Central Source. Vectors ripple inward (towards) this Source or outward (from it) towards peripheral sink. **With wave-like rings this model is simultaneously building a picture of spectrum, that is, scale of *energy* and *information*.** Natural Dialectic calls it the conscio-material spectrum or, as follows, gradient of creation.

Third is the idea of Mount Universe, the Cosmic Pyramid.

Although it also describes flow from Potential to impotence, a cone or, squarer, pyramid describes '*static*' hierarchy. Rather than a spectrum's continuity of difference this model discontinuously separates by grade. Hence, a useful representation is the stepped pyramid, also called a *ziggurat*. In this case each step of a **ziggurat** stands clearly for a phase, level or stage; and its capstone, a point that points beyond the finite grades below, implies peak infinity. **This peak is the highest point and source of what we call Mount Universe**.

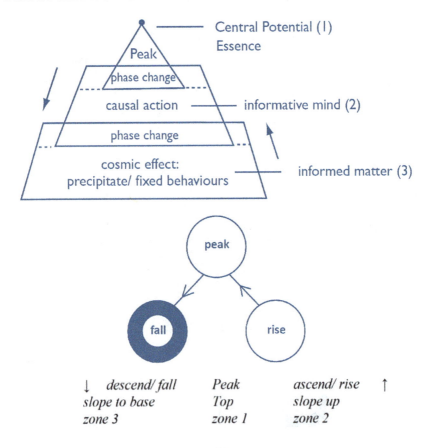

↓ *descend/ fall*	*Peak*	*ascend/ rise* ↑
slope to base	*Top*	*slope up*
zone 3	*zone 1*	*zone 2*

Models act as metaphors. Of these three you can use which one best illustrates a point you're making. Their Source, Centre, Axis, Peak and Pivot are identical.

And up and down (regarding motion or transformation) mark the basis of existence called, in other word, **duality**. Change and relativity are the nature of all things. But it is obvious, from the triplex stacks, that we are also dealing in **trinities**. The nature of this Central, Balancing Source is also, as we'll see, of great interest.

↓ (-) *negative   Neutral   positive*   (+) ↑

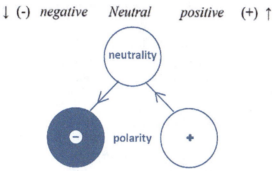

Neutral, positive and negative. The words in stacks are not arranged haphazardly. Their order represents a logic. Triplex by triplex, stack by stack we can and shall build a simple but a self-consistent picture of the world.

Moreover, three components of each model, <u>two vectors and a point of balance</u>, find various expression in each object and event in creation - even, we have already glimpsed, in description of creation as a whole. More subtly, they reflect operating principles whose permutations are expressed, in varying degree, as *tendencies* both psychological and physical. **Since the triplex nature of both models and stacks turns out to be all-pervasive to the extent of describing the three major tiers of Mount Universe itself, we call each member a <u>cosmic fundamental</u>.**[5]

If the triplex is naturally fundamental, it is also fundamental to the systematic operation of the Dialectic we are formulating and can now begin to use. While the orient has long worked and woven with these immaterial radicals, these fundamental threads, then western minds have not.

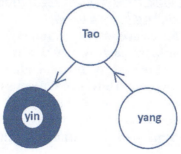

↓   *yin*   *Tao*   *yang*   ↑

You might be familiar with this far-eastern, Taoist triplet and its associations. Therefore, to head its stacks, Natural Dialectic uses an equivalent, equally ancient abstraction. **To better learn the basics of Natural Dialectic please consult Appendix 1 after which you can return to this point.**

Now we are in a better position to ask, **"How is all this related to science?"** [6] **How can stacks and cosmic fundamentals re-angle not the facts but the prism of our perspective?**

In fact, we can see and will elaborate the triplex logic within which we might profitably and comprehensively treat the scientific disciplines of (mind) psychology, matter (physics) and their connection, mind-with-matter, in biology.

> *A simple link from* <u>*psychology*</u> *describes the three basic conditions of information as (sat) potential informant (concentrate of consciousness) prior to (raj) active mind (both informant and informed) and (tam) passive, dormant or sub-conscious mind.*

> *A simple link from* <u>*physics*</u> *describes the three basic conditions of energy as (sat) potential prior to (raj) action and (tam) exhaustion.*

> *Thirdly, a simple link from* <u>*biology*</u> *describes the three basic conditions of life on earth as (sat) informative archetype, (raj) metabolism including a code carrier called DNA and (tam) finished product called developing or adult body.*

In other words, there exists a holistic, triplex (threefold) treatment of science as well as the one-tier materialistic one; and, as regards holistic logic's fundamental connectivity, there are thousands of possible stacks. I have used just an exemplary handful in this chapter. ***What they amount to is a system for generating fresh perspective, links you have not made before, new insights into cosmic connections.***

To summarise: **we have covered the 'grammar' of Natural Dialectic and are now over the hard part. Along the next 'plain-sailing' five chapters we'll see how three cosmic fundamentals fit with science and with the construction of Mount Universe. We can use a 'base-2' (binary) philosophical structure to gain a unified perspective regarding physics, psychology, biology and sociology; and there is a section in the Glossary, 'dialectical stack', designed to help students better understand and** practise composing their own frames of reference. These, if to-fro, binary logic is at its root, reflect the way our polar cosmos is constructed.

At this point we reiterate that, while **holism** includes an immaterial element, information, current science by default frames its explorations in a materialistic and, regarding life on earth, evolutionary framework. *Thus to reappraise the data in a holistic manner amounts to fundamental change in a basic concept of scientific discipline.* **This is what physicist and philosopher Thomas Kuhn called a <u>paradigm shift</u>.**[7]

After each chapter a set of questions (there are 170 in all) like the following will hopefully educe a better understanding of the structure of Natural Dialectic and how it reflects cosmos, that is, the orderly patterns of creation.

Questions 1

1. What is the difference between holism and materialism?

2. Is holistic 'Spin' simply an off-shoot of a venerable discipline called The Philosophy of Science?

3. Give an example of an immaterial element.

4. What, in brief, is Natural Dialectic?

5. What does materialism hold true concerning the origin of complex, codified biological form?

6. What three models are used by holistic Natural Dialectic to describe the structure of cosmos?

7. Where is non-conscious matter placed in ziggurat cosmology?

8. Where is it placed in the circular model?

9. Write a line of triplex 'stack' using the words 'sink', 'flow' and 'source'. Include vectors (arrows).

10. Which of these items is at the 'top'?

11. How does each of the three models symbolise 'source'?

12. Write a line of 'stack' involving process through time.

13. If material = physical then immaterial = ?

14. What are 'cosmic fundamentals'?

15. How might you express them in a triplex line of 'stack'?

16. What is the description of cosmos when thought of as a spectrum?

17. What two elements compose, in various proportions as on a sliding scale, this spectrum?

18. How is existence defined?

19. What is the holistic complementary opposite of existence?

    To answer the following questions you will need to check Appendix 1.

20. Of what does a Primary Stack consist?

21. Of what does a Secondary Stack consist?

22. How does any apparent contradiction, such as 'balanced motion', occur?

23. Give three examples of highly balanced motion.

24. Give an example of each in a less balanced state.

25. How might non-motion, potential or source be represented?

26. What is the opposite of Absolute?

27. The right-hand column of Primary Dialectic is a list of what?

28. And its left-hand column?

29. What does the Secondary Dialectic list?

30. Which fundamental qualities are used to divide each major scientific discipline into a triplex of treatments?

31. What's it all about? Name one fundamental question to which we might apply 'philosophical machinery' in the form of Natural Dialectic to obtain an answer.

You have now completed a challenging part of this course. *If you wish to complete your understanding of Natural Dialectic's simple grammar, please turn to Appendix 1.*

Chapter 2: Information

Let's straightaway, having outlined the structure of our philosophical vehicle, throw down the gauntlet. Let's enter the halls of scientific discipline announcing three great issues we'll address.

3 Great Issues of Science and Philosophy

Psychology: The Nature and Origin of Consciousness

Physics: Cosmogony, The Origin of Physicality

Biology: The Origin of Codified Life-Forms

In this exploration we draw a distinction between two basic modes of of enquiry, information gathering and processing. The first is **subjective focus** - contemplation which allows planning and perception of principles; the second is **outward focus** - perception by means of sensation, muscular manipulation and their technological extensions. Both modes will be discussed later in the section on psychology. For now, simply note that these *anti-parallel vectors* of mind correspond to what we call *bottom-up* and *top-down* directions.[8]

Bottom-up is, broadly, that of **the learner working from measurements of detail towards understanding patterns and principles by experience**. This kind of logic, **from detail to principle, is called inductive**. It is the way of naturalistic curiosity, that is, of experimental science.

Top-down is, broadly, that of **the expert working from principle and its application for making sense of details**. This kind of logic, **from principle to detail, is called deductive**. It is the way, from mind to matter, that holism works.

There is a third, important sort of logic that we'll come across. **It applies to unique historical events or futuristic speculation that we have neither seen nor can for sure repeat. It is called abductive reason or conclusion by best inference.**

Now, in dialectical respect of principle, I'd like to draw attention to duality within unity and the basic elements of the existential dipole drawn below.[9] We will shortly relate its triplex nature to Mount Universe. **First note that the two basic elements composing all the various forms of creation, that is, of existence are information and non-conscious energy.** In each exists a scale of informative and energetic phases we identify as, respectively, mind and matter?

Materialism, having taken immaterial information out of its equation, has decided matter is the basis of creation (though it does not specify the nature of the pre-physical nothingness that matter must have started from); thence mind is material excrescence from a brain and soul a pure (but physical) imagination. On the other hand, holism's revolution turns materialism on its head. The diagram below shows The Monopole, the nature of Sole Cosmic Independence - Infinite Essence; and its dependent, finite, existential dipole within which occur all motions of creation.

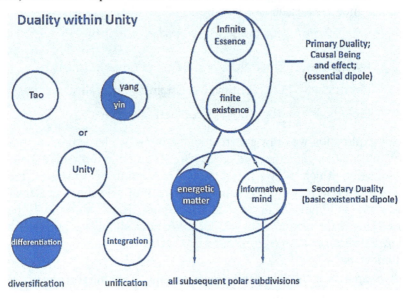

We are now in a position to relate cosmos in terms of **the triplex yin-yang-*Tao* model and the cosmic pyramid: (*Sat*) Essence with (*raj*) mind and (*tam*) matter.**

Cosmic Fundamentals and their Ziggurat
Three Tiers of Mount Universe with Sub-divisions

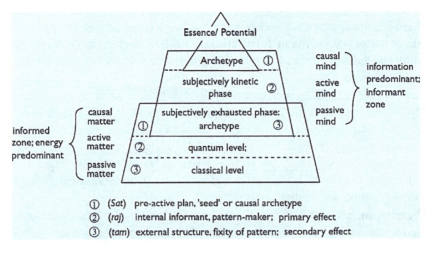

① (*Sat*) pre-active plan, 'seed' or causal archetype
② (*raj*) internal informant, pattern-maker; primary effect
③ (*tam*) external structure, fixity of pattern; secondary effect

tam/raj	Sat
existence	Essence/Potential
expression/creation	Pre-active Potential
appearance	Source
↓ lower	upper ↑
physic	metaphysic
energy	information
reflex/passive	active/creative
non-conscious	conscious
post-active phase	active phase
body	mind

This ziggurat, stack and the table below help to elaborate the previous diagram.[10] Together they make clear the triplex nature of cosmos in terms of the three cosmic fundamentals and the basic existential dipole. The whole is in three parts - (*Sat*) Source, (*raj*) mind and (*tam*) matter. The latter couple are themselves divisible into sets of three. This may be conceptually new to you but we'll deal with information in terms of its triplex sub-divisions later in this chapter and, similarly, with related scientific disciplines in chapters 3-6.

Upper Pole – Information
  (Sat) Potential Information
  (Raj) Active Information
  (Tam) Passive Information
Lower Pole – Energy
  (Sat) Potential Energy
  (Raj) Active Energy
  (Tam) Passive Energy

Before moving on, please note that 'potential' is used in a way not entirely correlated with that of physics. Also that the phase of *physical archetype* (see the above ziggurat and Glossary) is identified as a synonym of 'causal super-matter' or, simply, 'potential matter'. *One passive form of information is a memory.* Could archetypal memories[11] constitute the *potential* phase in the triplex subset of both physical *and* psychological levels? If so, then archetypal source becomes an important feature of holistic cosmos. For example, think of passive archetype as a 'seed program' that precedes yet governs the origin of physicality; so you might allow that its pre-physical operations followed different rules from those in physic's space and time. Could such program exist, like a kind of informative '*DNA*' in the holographic body of the universe, apparently nowhere - being metaphysical - but actually everywhere and every time at once? We'll check further in Chapter 4, the section on physics.

21

Every scale or hierarchy has extremes.[12] Now let's take a brief look at the ziggurat's extremities, that is, the thesis of its extreme initiation and, at the other end, consequent extreme antithesis. At base rests non-conscious matter, a universe whose ultimate sinks are perhaps the massive, local 'infinities' of black hole singularities. The opposite of non-conscious is conscious. In this case we identify, at immaterial super-state, the ultimate, uncaused source as the Singularity of Consciousness. This formless initiator is, logically, the extreme concentrate or purity of consciousness.

In other words, uncaused and eternal Transcendent Super-state is regarded as the Centre, Source or Pivot round which cosmos swings; and the sink/ periphery is non-conscious matter.

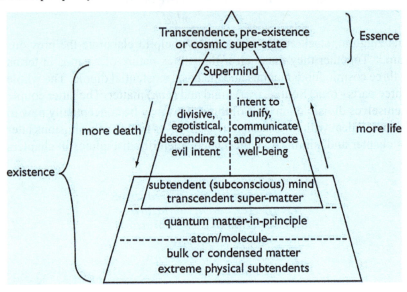

Why, though, put consciousness above non-consciousness? Materialism asks why *eternal matter* of some species shouldn't be the uncaused source of all there is. Science, though, at present believes that material cosmos had a start, all physical phenomena are offspring of first cause and steady-state theories of eternal matter have been binned.[13]

Conversely, why should Super-state hierarchically precede mind as well as matter? Why should metaphysic more likely than everlasting, physical energy be the uncaused Cause of Everything?

It is true, *top-down*, that products subtend (are of lower grade than) their producers. A producer stimulates and stirs things up. Cause precedes consequence. *From this perspective active informant precedes its passive, informed consequence, metaphysical mind plans physical arrangement and super-natural potential precedes its possible orders. **Mind first, body after; body's an appendage of its mind.*** It happens all the time with us: idea (or desire) precedes outcome. *But doesn't it, that physical depends on metaphysical, invert your 'normal' sense of things?*

So, for example, we say that sleep subtends waking, sub-conscious subtends conscious. Conversely, waking transcends (is of higher grade than) sleep. And, as this and Chapter 3 will demonstrate, you transcend Secondary to reach the qualities of Primary Dialectic. Various states are like the bands of a spectrum - ultra-violet, visible (our waking state) to infra-red and black. **We're emphasising the notion of _hierarchy_ central to a tiered universe. Spectrum illustrates an idea of hierarchy, scale or, in ancient parlance, Jacob's ladder.**

Hierarchies always have source. A boss. An activator or a cause. *So now the hierarchical interest turns to causes.* Of course, we can think of causes in terms of knock-on effects. They impact, in physics, chemistry and psychological reflexes, in '*horizontal*' chains. **No doubt, energetic causes push effects; their arrow, physic's arrow, runs from past cause to present effect; in this sense they 'bump you from behind'.** And 'bumping' loses energy. Things suffer (though you'd hardly think it as regards a proton or electron) from increasing weariness called entropy (see Glossary). They run out of steam. We call such universal physic horizontal or energetic causation.

What, though, about a cause that is conceptually implanted? What about not energetic but *informative causation*? **This is goal-oriented; and goals are in the future pulling you their way. They pull you towards your chosen future; they lift forward. They are metaphysical *attractors*, guides that govern your behaviour. Purpose *leads* you through the world.** Not material but immaterial, such leadership is not by force of gravity, electric charge or magnetism; information is metaphysical not physical; it's psychological and, although at this point you may want to know exactly how, be patient - as the course elaborates we shall come to see. **In short, mind is purposeful. It strives to unify you with your wishes and thus metaphysic's arrow flies, from future back to present, anti-parallel to physic's.** Thus you are guided, present to the future, by your plans or programs realising goals. For example, take the rules of football - unseen and unexpressed till players play a game.

Waggle your little finger. What *causes* what? The line of operational command runs from conscious, informant mind through sub-conscious templates (memories, instincts and so on, dealt with later in Chapter 3) over a psychosomatic border to innervated, muscular body. Between the zones of purpose (mind) and non-purpose (body), incoming or outgoing information would be first translated to, or last translated from, the chemo-physical side by the product of 'excited' electrons, that is, electromagnetism. Where *matter subtends mind*, this electronic phase is in turn subtended by the biochemical; and both levels occur within bulk, biological structures, that is, cells, nervous system, muscles and a coherent whole body. **This places your skin-and-bone finger-waggle at the base of an informative/ energetic hierarchy.**

We thus call this process <u>vertical or informant causation.</u>[14] **It is the order of an act of creation.** An idea is developed. Its 'phased intent' drops from creative mind to created material form. As a later chapter will elaborate, your own sensible, physical form (perhaps including the origin of its shape) resides at the bottom of such a hierarchy.

The Order of an Act of Creation

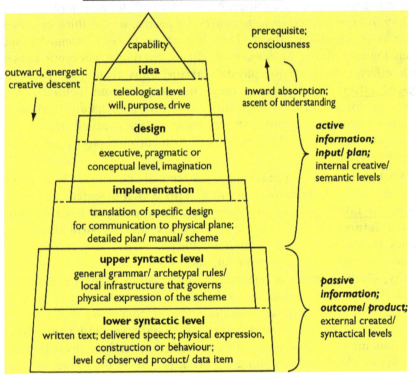

In the previous chapter we learnt how to read a full (Primary and secondary) stack from top to bottom; and in this chapter it was noted that start (or source) through mind to finish (or material end-product) can be portrayed as three tiers of Mount Universe.

We are now in a position to elaborate this 'reading routine' to simply define, using two full stacks, <u>the order of an act of creation</u> as regards cosmos itself (cosmogony with its product, studied by cosmologists) and any single, individual event within it. At the same time these Essential and existential routines demonstrate both the hierarchical and cyclical nature of creation.

It is useful to rephrase such a new and powerful routine. The logic of Natural Dialectic indicates a drop between immaterial and material extremes, that is, from pure concentrate of consciousness to pure non-consciousness (the state of matter). It thereby suggests a 'conscio-material gradient' of cosmos, that is, a spectrum or hierarchical structure based on level of consciousness. In this hierarchy Immaterial First Cause is the Zero-Point; it is the Origin, a Source called Essence.

Such Universal Initiation, giving rise to the cycles of existence, is illustrated in the first of the next two diagrams.[15]

Essential Cosmos

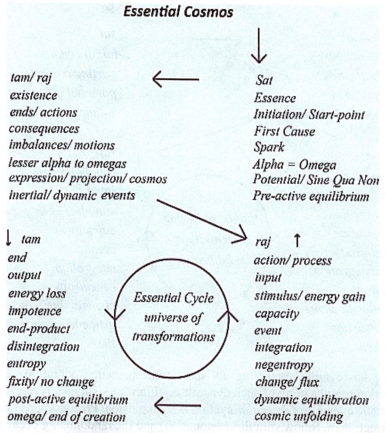

tam/ raj	Sat
existence	Essence
ends/ actions	Initiation/ Start-point
consequences	First Cause
imbalances/ motions	Spark
lesser alpha to omegas	Alpha = Omega
expression/ projection/ cosmos	Potential/ Sine Qua Non
inertial/ dynamic events	Pre-active equilibrium

↓ *tam* | *raj* ↑
end | action/ process
output | input
energy loss | stimulus/ energy gain
impotence | capacity
end-product | event
disintegration | integration
entropy | negentropy
fixity/ no change | change/ flux
post-active equilibrium | dynamic equilibration
omega/ end of creation | cosmic unfolding

Essential Cycle universe of transformations

There exists, as well as Essential, existential initiation. This simply means a start-to-finish of every single event in a continually mobile cosmos. This case is illustrated in the second diagram called 'existential event'.

Every event cycle, succinctly expressed as causal stimulation yielding action that dies away to an end-point, is recycled with a fresh stimulus. In this, the cause (or potential) for any event may be vertical (psychological) or horizontal (by physical reflex or knock-on effect). In other words, the second illustration, existential cycles, includes the vertical causation or psychological events of purpose/ desire; and the horizontal causation of physical events, that is, mindless, reflex cycles of interaction ('passive information' in the yellow illustration).

Events can happen simultaneously within each other e.g. a cloud in the sky or atomic oscillation within molecule within larger body. It is not nature's but an observer's problem which event is singled out for focus.

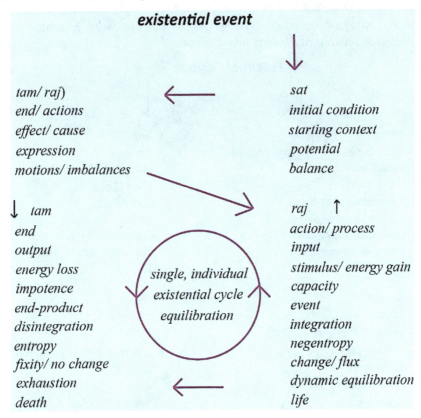

existential event

tam/ raj)	sat
end/ actions	initial condition
effect/ cause	starting context
expression	potential
motions/ imbalances	balance

↓ tam	raj ↑
end	action/ process
output	input
energy loss	stimulus/ energy gain
impotence	capacity
end-product	event
disintegration	integration
entropy	negentropy
fixity/ no change	change/ flux
exhaustion	dynamic equilibration
death	life

single, individual existential cycle equilibration

Regular cycling/ recycling in space-time is called vibration. The oscillation round a norm of such dynamic equilibrium is called a wave. And a dramatic example of *irregular* cycling is an explosion. Preconditions must be correct; there follow stimulus (detonation) and the randomising decay of fall-out until motionless silence, the end.

Put another way, existential cycles, whether psychological or physical, amount to causes and effects within the regulated context of animate and inanimate conditions. They may be mindless and automatic or the outcome of purposeful intent. For example, a machine (see Glossary) with cog-wheels systematically coordinated to produce a desired operation (such as a clockwork watch) is a specifically designed outcome. A clock is many orders less complex than any bio-machine, that is, biological body. Whichever way, mindless or mindful, many other kinds of cycle such as planetary, temporal, ecological and bio-cybernetic are familiar to everyone. So are life-cycles and, from idea to fulfilment, the cycle of desire. Existence runs around in circles.

As you have seen the holistic view holds vertical cause *higher* than its effect. The finger-waggle demonstrated how you personally operate. However, we can reach beyond the normal cycles of existence both psychological and physical. We now focus on Source, Centre or Highest, All-Informant First Cause.

Aristotle believed in a First Mover itself unmoved by any cause. St. Augustine observed that no 'efficient cause' can cause and thus precede itself. Thus causal order can't be infinite; there needs to be an uncaused primal cause, a First Mover. Existence is composed of caused, finite events. Whatever begins to exist, asserted the *sufi* Algazel, has a cause; *and 'something which begins has a sufficient cause' is also the modern principle of causality*. **This principle is constantly verified and never falsified. The physical universe began to exist and therefore has a cause**. What is caused is not eternal. It is finite. Its effect becomes a further cause. Thus all existence is a changeful network made of causes and effects; creation is an action and reaction zone.

In summary, it's as simple as it's crystal clear. What starts to exist is always caused. We presume the physical universe started and eternal matter, that is, endless change, is not a feasibility. Therefore, material cosmos, known as nature to the natural sciences, started to exist and has a Natural Cause. *This cause, preceding physicality, is physical non-being.* It is nothing physical but causes cosmological effects. What came (or comes) before material creation's energy, space and time must itself be time-less, space-less, immaterial. Preceding its secondary causes and effects the primary, first cause must be physically uncaused; it is uncreated in a naturalistic sense, thus super-natural. Metaphysical.

What, therefore, is the Nature of Informant Metaphysic?

Primary (Metaphysical) and Secondary (Physical) First Causes[16]

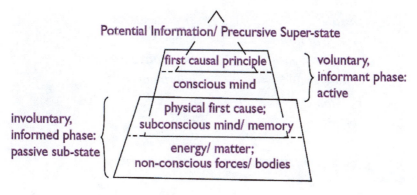

The first thing to note is that, in cosmic terms, *physical first cause is **not** the same* as *Psychological First Cause* (or Pre-motive, Essential Super-state). In short, the following ziggurat and causal cones illustrate two very different kinds of first cause, the **higher conscious** and the **lower unconscious**. Their natures will be elaborated in the following chapters.

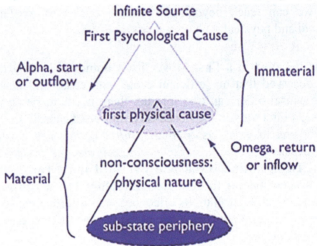

Now, after this preamble, it is time to focus on the nature of Natural Dialectic's immaterial element, information.[17] We are information junkies. We want the news, crave more information and want to know everything! Perhaps not only externally in the universe (including our bodies) but on the immaterial, contemplative mind side too. So what is info? It is defined as facts or fictions learned, knowledge, understanding or 'intelligence' (e.g. espionage or signals intelligence). It is also computer data or what is conveyed by a particular arrangement of things or symbols. The arrangement of symbols, representations or stand-fors is called code (see Glossary). Code is language. **But is there *any* kind of information without mind?**

tam/ raj	*Sat*
below	*Transcendence*
range of forms	*Super-State*
output/ event	*Pre-motive Potential*
matter/ mind	*Informant*
↓ *tam*	*raj* ↑
passive	*active*
matter	*mind*
passive information	*active information*

Top-down, *to receive and signal information is the immaterial gift of mind and never matter.* **It is irreducible to scientific scrutiny.** Pick up a postcard, menu, letter - anything informing you. The object is reducible to chemistry and physics but the *sign* or *message* it conveys is not. Signs and signals always have a purpose; objects *per se* never do. In other words, information is fundamental to our being, but does not fall, in a semantic sense, within the scientific remit.

Particularly, from this perspective, a chance-based, knock-on, mindless origin of any body's bio-programs might seem an irrational hypothesis. Why? In the case of information transfer code is as necessary as the symbols employed. **Comprising the basis of all technological and bio-systems, code constitutes a mental system.** Thus, if it is found in any system it can be deduced that this system originated in concept. Accordingly, genetic code, including *DNA* and protein components, did not arise by chance but from concept (see the rest of this Chapter and Chapter 5 *passim*).

Bottom-up, **however, everything is seen as energetic interactions. Energy is the oblivious 'informant'.** Such information, even in the case of brain, must be elicited by chance and natural law. **Physic generates its incidents but never has a goal; locked in 'horizontal', reflex causal cycles matter thoughtlessly churns transformations.** So, code or no code, happenstance is at the root of everything. Is such materialistic (and yet also metaphysical) position an illusion, a confusion or the truth?

Firstly, in this dilemma of perspective, let's distinguish everyday but meaningful sense perception from objective scientific analysis. *In mathematics of the latter sense 'information' must not be confused with 'meaning'.* So what exactly, in this apparent shortfall, *does* the word mean?

Claude Shannon

Norbert Wiener

It was Claude Shannon who, with Warren Weaver, first devised a mathematical and thereby scientifically acceptable definition of information. Shannon treated its transmission and storage in purely physical terms according to statistical formulation. In such transactions his unit was the *bi*nary digi*t*. Whether by electrical or quantum computing, this on/ off, one-zero 'bit' allows the quantitative properties of strings of symbols to be calculated. His theory inversely relates information and uncertainty. The more uncertain, the less probable a sequence of symbols or arrangement of materials is, the more information it is calculated to contain. Rephrased, an amount of information is inversely proportional to the probability of its occurrence by chance. Simple, repetitive or predictable sequences contain less and complex, irregular arrangements more information. **Thus 'Shannon information' is simply a measure of improbability.** *It makes no judgement whether such irregularity is specified; it involves no sense of meaning.*

ZNQW&NSIXT AZ2NVB
COMPREHENSIBILITY

Shannon's definition is suitable for describing statistical aspects of information such as quantitative aspects of language that depend on frequencies (e.g. as how many times the letter 'a' or the word 'and' occurs) but it treats any random sequence of symbols as information without regard to its concept, meaning or purpose. In other words, the more improbable any arrangement the richer is its Shannon-defined information content. *In short, two messages, one meaningful and the other nonsense, can be exactly equivalent according to this form of analysis.* For example, **ZNQW&RSIXT AZ2HVB** and **COMPREHENSIBILITY** are assigned the same value. In other words, Shannon has reduced information to a statistical quantity; he has shaved off any sense of meaning, cut out sense of purpose in numerical analysis. **Shannon-shaving deals the same with purposive (vertically caused) and non-purposive (horizontally caused) complexity.**

In this mode of thinking a 20-letter randomly-generated and meaningless sequence contains information richer than the simple phrase 'I love you'.

It is important to grasp how Shannon, thoroughly negating logical reality, accords randomness and purpose equal status. **But randomness is really reason-in-reverse; it is information's opposite. So when it comes to language or to mechanism, including those embodying biology, <u>such conflation is an error of first order</u>.** Shannon's analysis does not distinguish between presence of mind, authorship or creativity and their absence; it fails to recognise purposive specificity; nor does it accord function or meaning any premium. *Thus order and precise meaning - usually complex, always accurately coded - might as well be able to emerge from randomness or senseless motion under natural laws of physics.*

Shannon's definition involves counting the frequencies of physical events; and physicists, measuring non-conscious energies, sometimes refer to 'information-carrying' interactions. However information is, like energy/matter, a fundamental quality as well as quantity. Science investigates the

passive, lower syntactical phase of information whose causes inform horizontally. Such energetic interactions are mindless; the only information involved is, extricating principles and patterns, in an observer's mind. Ignoring this, however, you may fool yourself and think in topsy-turvy terms of 'senseless design', an anti-teleology well-known as Darwinian evolution (Chapter 5).[18]

Informative capacity inhabits the *arrangement* of material - not just any old arrangement but one with meaning that serves purpose or accords with principle. For example, everything around you in your room, not least the simple chair you are sitting on, involves informative capacity. It involves the **active** quality of information, mind. Purpose weighs no more than understanding; both weigh just as much as meaning. In the scientific balance meaning weighs as much as abstract theory. Each is lighter than a feather. Universal mind would weigh precisely nothing yet remain a vital hidden variable. *If, though, information is absent massively, what is it?*

If immaterial then, this volume shows, information renders materialism with scientism (but not science) most illogical! Indeed, Norbert Wiener, mathematician and founder of cybernetics and information theory, said, *"Information is information, neither matter nor energy. Any materialism which disregards this will not survive one day."*

> Norbert Wiener, mathematician and founder of cybernetics and information theory, said *"Information is information, neither matter nor energy. Any materialism which disregards this will not survive one day."*
>
> Information is not a property of matter. Purely material processes, unguided except by natural law, are fundamentally precluded as *sources* of information. That is to say, there is known neither law nor process nor sequence of events through which oblivious matter can create or collect information.

There is, furthermore, known neither law nor process nor sequence of events through which oblivious matter can create or collect information. **Information is not, therefore, a property of matter.** Purely material processes, unguided except by natural law, are fundamentally precluded as *sources* of information. Information is not a thing itself but a representation of physical things and metaphysical entities. Physical nature wears no numbers, bodies are devoid of words; these are planted or taken away by mind.[19]

To rephrase, information involves the transformations and arrangement of material phenomena but such phenomena *per se* are purposeless; they can inform a mind (by sense, mathematical descriptions etc.) but not their own aimless, mindless, non-conscious oblivion. However, although the default materialistic position is to claim laws of nature operate within a field of chance encounters, it is possible to holistically argue that physic provides a dynamic stage for purpose, that is, for life. *There may be a purposive source (or program) that informs specific, automatic behaviours.*

In this case, let's take a second angle on the way that cosmos and the human individual are informed.

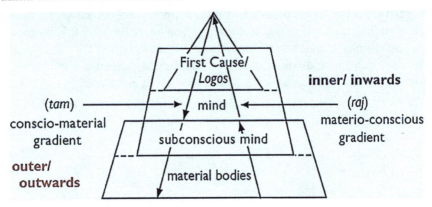

This is a key illustration. Its ziggurat reflects **the information gradient from active, inner to passive, outer phase;** and also the diagram showing 'Two Kinds of First Cause'. As such, it illustrates the anti-parallels of a dynamic conscio-material gradient that run up-in or down-out. This drops from Clear Formlessness at peak through psychological and physical forms of various quality. The latter are trapped in reflex, oblivious cycles; the former involve, at least for humans, a measure of choice.

For '**inside information**' you can see the way to go; and how, conversely, sensation leads you 'down' and 'out' into the world of '**outside information**', that is, of physical events. You can thereby easily understand why the sage turns his contemplative focus inwards towards unification with First Cause; and a scientist outwards to the details and, in principle, unification of an essentially non-conscious universe. Sharpened focus. Concentration does the trick. One man could incorporate both sage and scientist!

Check the diagram below. X, where you live, is the tipping point between the inner and the outer worlds. X, the 'third eye' of thought, is your micro-cosmological axis. Called mind, it is the information centre where you live. Now, with regard to macro-cosm too, you have your **cosmological bearings**.

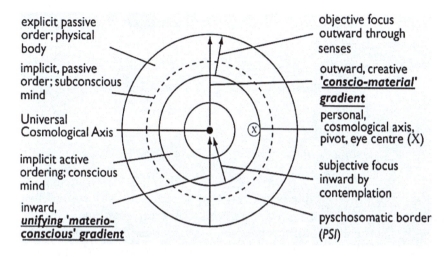

explicit passive order; physical body

implicit, passive order; subconscious mind

Universal Cosmological Axis

implicit active ordering; conscious mind

inward, *unifying 'materio-conscious' gradient*

objective focus outward through senses

outward, creative *'conscio-material' gradient*

personal, cosmological axis, pivot, eye centre (X)

subjective focus inward by contemplation

pyschosomatic border (PSI)

Where peak is central axis, the diagram below is another way to understand the previous key illustration. *And if we refer back to an illustration called The Order of an Act of Creation we can combine all three models to demonstrate, as follows, the triplex nature of information according to the fundamentals sat, raj and tam.*[20]

First in this triplex issue comes *(Sat)* **Potential Information** at the central cosmological axis or source of mind. Such conscious, inside information is what the contemplative sage is fundamentally after. Look *top right* where a sage's hand is pointing. It leads up towards Absolute, as opposed to relative, Truth (see also Chapter 6). It is experienced as Illumination; and Axial Illumination, projected back down into relativity, becomes First Cause Psychological (see Glossary and Index).

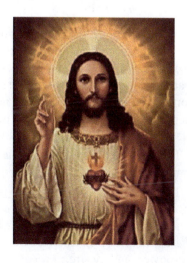

changeless	Illumination
mind	Essential State
relativity	First Cause/ Logos
relative impotence	Potency/ Control
↓ unconscious sub-state	mental spectrum ↑
darkness	range of shades
total lack of awareness	degrees of knowledge
mind dormant	mind awake

(Raj) active information **is only created in conscious, choice-flexible mind.** Its subjective experience knows, feels, observes, learns, understands and wills. *In the yellow illustration we saw what, for your microcosmic part, constitutes the order of your personal acts of creation. Could this order of vertical causation reflect a macrocosmic, that is, universal pattern?*

tam/ raj	Sat
expression	Idea/ Purpose
↓ tam	raj ↑
passive element	active element
informed	informant
physical elaboration	conceptual elaboration
created outcome	creativity
output	input
end-product	development inc. manufacture
hardware	software
machine	program

Idea is the seed. There accompanies an intention to develop this embryonic inspiration's possibilities. Development of idea, grand or trivial in scope, is the basis of all purposes. Can we further understand the order of an act of creation from the way we seek to elaborate, that is, express an idea? From trivial creation of finger-wagging up to creation of a work of art or science all involves, essentially, the same order of process or, as already noted in that illustration, **phased intent**.

The outcome may be simple (as in finger-wagging) or complex. Complexity itself is of two kinds.

The first, **purposive complexity**, is psychological. *It is a function of information gain, expansion of consciousness and a capacity to grasp principles and purposely, creatively exploit the principles and possibilities inherent in any circumstance.* Conscious creativity is filled by information, will-power and desire. *Indeed, will and desire are the psychological*

34

equivalent of physical electromagnetic radiation. Will is like electricity, desire like magnetism. Volitio-attractive; pushy will and pulling of desire. These are the complementary energies of mind.

An increasingly concentrated focus of attention wakens to greater capacity, flexibility and possibilities for specifically ordered, coherent or *active complexity*. *Such purposeful complexity* works against the 'downward' wear and tear of time and chance; it codifies and specifies design - which chance cannot. It is an instrument of biological survival, intellectual enquiry, technology and artistic creation. We continually experience it. Its proof, in artefacts and actions, pervades our lives.

Transmission of purpose involves, in the outward direction, imagining and planning its realisation; and, on the inward, deciphering another's plan. This is active teleology. In fact, whenever an intention is physically expressed it works through mechanisms and machines. Machines, which include biological bodies, operate according to physical law and fulfil their function using, in one form or another, energy. They specifically accord with plan and inform the world in ways unguided nature can't. As such they obviously link information (that is metaphysical) with energy (that's not).

Such causal reason generates meaning. *All active information is meaningful.* Implementation of a plan involves *semantics whose specific meaning transcends the generality of grammar and syntax.* The latter are simply vehicles of reasonable expression; their immaterial symbols are needed to make connection with the material world and thereby order it. They translate mind to matter. Coding and decoding, using speech-through-air, written word or other forms of signal, are core semantic business.[21]

Codes make sense. *Meaning* **and reason are the important, active ingredients behind data transmission and storage. Nevertheless these passive instruments of communication are important. We need frameworks (hardware) within which to manage information (software).**

Information's Infrastructure - Code

↓ *irrationality*	*logic* ↑
no-message	*message*
no-code	*code*
physico-chemical maelstrom	*psychological scheme*
accident	*teleology/ purpose*
chance construction	*technology*
mindlessness	*mindfulness*

A command erases randomness; it is a deliberate restriction imposed to cause a non-random outcome. It employs an agent of restriction - sign, symbol, code of one sort or another. *This is the world of signals, semantics.* *What is spoken is not a matter of chance.* **Whatever is encoded is intentional**. Code and incoherent chance are chalk and cheese. They never mix. Language, other symbols and the construction or decipherment of meaningful communication (called semantics) make up the third level of information. **In other words, this is the level at which ideas are framed in code or blueprint before their presentation in material form.** *It is important when considering biology, whose basis - metaphysical basis - is code.*

On the other hand, *(tam)* **passive information is the expression, external to conscious, flexible mind, of active information.** Such objective, energetic forms may interact or be dynamically exchanged (as, for example, in the case of an impact) but are, being mindless and therefore wholly imperceptive, meaningless. This type of impression is effectively a record stored either in subconscious mind (featuring relatively inflexible memory, instinct and so on) or using matter (where instruction is carried by arrangement on chemicals such as clay, papyrus, ink, *DNA* or other messengers). Storage may be fixed (as in a file or photograph) or dynamic (as in a running film, program or automated mechanism).

Note that the second, **non-purposive** kind of **complexity** is the product of reflex forces and particles; oblivious, physical reactivity creates aggregations such as stars, mountains or snowflakes. Such complication is mindless, aimless, *passive*, purposeless; and its universe could not create a cup of sweet tea in a billion years.

To cut the story short, *the upper level of code is its abstract infrastructure, the lower is its physical arrangement (in chemical or energetic e.g. vocal form). Metaphysical precedes physical.*

Top-down, therefore, chance loses out. **Code is always the result of mental process.** Its meaning is its message; this message, which communicates a reason or anticipates an outcome, thus has purpose. How can you, except by twisting logic backwards (as evolutionary naturalism has to do), have a code without conception of its product? Because irrational oblivion has no goal; it can never specify a reasonable plan or a machine.

If a basic code is found in any system, you can always trace its source to conscious intelligence - more easily if that code is optimised according to such criteria as ease and accuracy of transmission, maximum storage density and efficiency of carriage (such as electrical, chemical, magnetic, olfactory, on paper, on tape, broadcast etc.) to its recipient; and if, above all, it works and orderly instructions are unerringly responded to.

A coder takes no chance. Randomness is eliminated. Message constrains recipient to react in a certain way. In mind response is actively chosen; in

36

reflex matter the response is passive, automatic. Thus, with a computer, compiler and processor are mindless mechanisms but a programmer is not. He determines the code and its operation: error is rigorously debugged. By definition, mistake or randomness degrades his work; and the job of any editor is to eliminate interference, 'noise' or mistake.

The rules that govern physics' fixed, non-conscious program (Chapter 4) are immune to chance and change; yet physical behaviours are prone to both. In the case of code, however, chance neither creates nor transmits information. On the contrary, accidental interference always (unless accidentally reversing a previous degradation) degrades meaning and, by degree, renders information unintelligible.

Moreover, if information is immaterial then a frame (such as the holistic one that I propose) is needed; the metaphysical needs be included along with the shock from subsequent theories of Source, Origin or Potential to materialistic physics, psychology and biology.

All active information (whose psychology we'll address later) is meaningful. **Implementation** of a plan involves *semantics whose specific meaning transcends the generality of grammar and syntax.* **We need frameworks within which to manage flows of data.**

Code is devised and stored in mind but information's physical expression obviously employs material arrangements. The alphabets of such code include, as well as humanly devised systems of communication, certain bio-chemicals (supremely, *DNA*), vibrations, binary nervous code and the sub-atomic elements, forces and atoms of physics and chemistry. The former couple are specific to life forms - possibly, at least with respect to *DNA*, anywhere life may exist in cosmos. The latter also constitute a universal code. This code, immune to human manipulation, dictates, through the agents that a study of phenomena elucidates, the way things naturally turn out. *Looked at this way cosmos is a Grand, Dynamic and Encoded Text.*

The lowest, physical level has, therefore, already been explained. **This is the part that bottoms out in light, sound, fluid patterns and in crystalline solidity. It is the 'external', 'objective' or physical side of the *PSI* (see Glossary); and, while forces and atomic particles comprise its primary expression, the hard, bulk universe that we survey is its secondary.** *At this level we find data items, that is, materials whose arrangement is wholly reflex or automatic.* Finally, therefore, let's examine three lowest-level but still complex expressions of purpose. Verbal, musical and mathematical codes have produced the glories of humanity.

Music.

A musician brushes sound on silence; at his signal from thin air the colours of a song appear.

The old word for integrated order is harmony.[22]

Harmony is balanced. Health is balanced. Both are forms of dynamic equilibrium. Balance is the key. Harmony is an archetypal formulation of energy, constrained only by its type of instrument, harmonics and the skill of its composer and musicians. Melody is a profound form of information-in-motion. It is, like light, a medium for the vibratory transmission of meaning, one that unlike speech is universal. And The Meaning of Illumination? Sound and Light of *Logos* issue from creation's Source; *Om*'s vibration orders and sustains the composition of our universe. What more Natural than the Heart of Nature?

In this case Metaphysical First Cause does not need numbers, words or brain! Logical Archetype, primordial vibration's song, alone will do! Symphony composed by *resonant association* is enough! The vibratory energy of First Cause and its subsequent constructions would embody the internal logic, patterns or rhythms that pulse through each level of the grid of creation; and from the flux of vibrance issues form (see Chapter 3: Cymatics). Music well expresses operation of the three cosmic fundamentals (Chapter 1), two of whose oscillations circulate around their central point of balance. Who is surprised that attractive harmonies of music are the food of love, bliss and healing? Healing physical as well as psychological? **Therefore, beyond the other cosmic models and because it can't be posted in a diagram and is best left to great composers, music is Natural Dialectic's Master Model.**

Machine.[23]

The table you are sitting at is a device of simple and unmoving parts. Its construction serves a purpose and, therefore, has 'mind in it'. Its shape is 'passively informed' from chemicals that would never take that shape without direction. In fact, every organism marks the world with trivial, crude or intricate constructions for survival and, for some, passion's further purposes. Look around. This room is everywhere imbued with mind. Physical and chemical analysis, howsoever exhaustive, will never of itself find mind and, therefore, reason in its material parts. **The hidden factor in plain sight is information.** Everywhere you easily detect informants and informed. Why, therefore, is information 'outside science' just because it's metaphysical not physical?

"Machines", argued philosopher/ scientist Michael Polyani, "are irreducible to physics and chemistry." They are irreducible because they involve immaterial purpose, the stepwise development of a plan of implementation, a directed cohesion of working parts and, of course, the thoroughly non-material anticipation of an operational outcome.

Let us rephrase. *A machine involves a system of well-matching, interacting parts that, unless any is removed or degraded, contribute to a function or produce a targeted result. Such systems are therefore specifically and irreducibly complex; <u>and so to work they must be made at once</u>.* In this case which comes first - a machine or its concept, a work of art or its inspiration, the chicken or its egg?

<u>**Let us finally be absolutely clear that, although machines are operationally subject to the constraints of natural forces and environmental context, they are never created by them.**</u> *No machine ever appeared as a consequence of the addition of random free energy into an existing system.* **Physical nature can't create machines; yet information and machinery are closely twined in living and, indeed, all teleological constructions.** *You want evidence for holism? This is fact.*

Mind Machine.

Mind is a symbolic representer and recorder. A computer is a mind machine. Inspect the logic of its functionality. Examine integrated circuits. Their molecules, like those of a brain, show no sign at all of mind. They are not even biochemical but metal, plastic, silicon and so on. Yet they are replete with passive, rigidified order. In this respect each one's determination cries out its ghost, the active order of its maker. **The whole machine is absolutely full of maker's information.**

Programs, coded programs are a mind machine's intent. **They express the will that mind invested in machine.** *It needs be re-emphasized - mind seeds the development of machine. Every machine (including body biological) expresses the ingenuity of its maker.* Not in its atoms *per se* but in its purpose, design and lawful operation. Machines passively embody information. Mind-machine information is passive. Active has produced passive information. *There is, to conclude, known neither law nor process of nature by which matter originates information.* **Information is the immaterial element.**

Questions 2

1.   What are the three great issues of science and philosophy?

2.   What does 'top-down' mean?

3.   What is inductive reasoning?

4.   And abductive reasoning?

5.   How does abductive reasoning impact science?

6.   What is the basic binary dipole (or duality)?

7.   What is the basic binary duality of existence?

8.   What are the three main phases/ bands of cosmos?

9.   Each of the poles can be subdivided into its own triplex. What are the three main sub-levels of the energetic pole?

10.   What does the 'finger-waggle' illustrate?

11.   In brief, what is an archetype?

12.   Top and base are diametrically opposed extremes. Every band, grade or level has its highest and lowest part. However, what are these two extremes, the peak and peripheral nadir of the *whole* conscio-material spectrum, called?

13.   Can you suggest candidates for each extreme?

14.   Does the order of an act of creation apply to you or to cosmos as a whole?

15.   What is vertical causation?

16.   Give an example of vertical causation.

17.   What did Norbert Wiener say about information?

18.   What is *potential* information?

19.   Define *active* information.

20.   Define *passive* information.

21.   Why does Shannon's work have nothing to do with 'active information'?

22.   What is the difference between purposive and non-purposive complexity?

23.   What is *always* the source of code? Could the non-conscious, material cosmos ever generate signals?

24.   Why are codes, plans, mechanisms and machines "irreducible to physics and chemistry"?

25.   How is a computer a mind machine?

Chapter 3: Psychology

Bear in mind, as we approach psychology, that Natural Dialectical Philosophy *is* internally self-consistent. Let's equally and importantly remember that, as in the finger-waggle, its knock-on networks of causation are not simply 'horizontal' (as with one-tiered materialism) but its 3-dimensional structure includes 'vertical' causation.

Lao Tse is seen here with Swiss psychologist Carl Jung. **Tse's cosmic fundamentals and the basic form of Natural Dialectic (the dynamic interplay of complementary opposites) are illustrated below.** Jung was deeply influenced by the Taoist I Ching or Book of Changes.

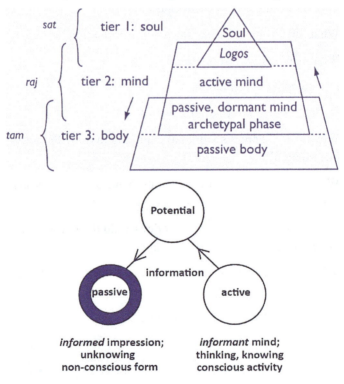

tier 3/ tier 2	Tier 1
passive/ active information	Potential Information
body/ mind	Psyche/ Soul
lesser selves	Transcendent Self
material/ mental forms	Conciousness
↓ physic	metaphysic ↑
energetic	informative
passive information	active information
non-conscious	states/ grades of mind
body/ tier 3	mind/ tier 2

With the idea of a 3-tiered universe do you also recall the logic of a contemplative monk or yogi - man, microcosm of the macrocosm?

In meditation they seek to remove the existential motions that arise in body (matter) and in mind. Once the waves of mind are tranquillised they no longer obscure *Psyche* (Transcendent, Pre-active Truth).

motion	Essence
finity	Infinity
relativity	Absolution
appearance	Truth

This Informant Subject[24] forever precedes the forms it predicates, forms composed of you and every other thing. Infinite projects the finite. Essence both transcends and yet composes its existence, that is, informs its programs and substantiates the energetic patterns of creation. This conditionless condition is, therefore, Supremely Paradoxical. Being of Potential Information communion with which was, in the previous chapter, identified as every mystic's goal.

> *Such Supreme Paradox is, beyond existence,*
> *Absolutely Non-Sensible; absurd as any quantum*
> *trait it is, however, furthest from absurd; it is our*
> *foundation and our starting-point.*

First Motion, first restriction of the Infinite is called First Cause, Causal Mind or Super-Conscious Archetype.[25] The previous chapter noted many labels; here the ziggurat locates it as *Logos*. It also noted that, as well as primary metaphysical, there is also secondary physical first cause. Upper and lower initiators. The latter, unconscious archetypal phase (tier 3) features in the previous, this and the next two chapters.

However, this is all *materialistic nonsense.* Claptrap.[26] So you have never met such notions in the context of a scientific course. **Materialism deems that thought and subjective experience (by sense or otherwise) are things**. But could senseless atoms ever yield their own experience? Or mathematics? Reasoning? Morality? Intelligence?

Remember that the world, for a materialist, is basically colliding particles and interactions due to force. *And, biologically, neither cells nor their nucleic acids know or care a fig.* Such oblivion is ignorant of laws of logic, creativity or purpose. Why should compound intricacies of atoms, up against the laws of entropy, codify for algorithms and the mechanisms for an intricate survival? Even *if* blind carelessness developed nerves (a great guess in doctrinal dark) why should their chemical reactions produce a calculation or a sonnet? Physicist **Sir John Polkinghorne** rams the point home. '**Thought is replaced by electro-chemical neural events. Two such events cannot confront each other in rational discourse...The very assertions of the reductionist himself are nothing but blips in the neural network of his brain. The world of rational discourse dissolves into the absurd chatter of firing synapses.**'

Please, therefore, allow at least the axiom that information, active or passive, might exist in immaterial form; and that mind is metaphysical/immaterial.

The word *'psychology'* means study of *psyche* (from a Greek word generally translated 'soul').[27] But current 'psychology' were better termed *'noology'*. This is because the discipline studies mind, for which the Greek is *noos* (from the verb 'to notice' or 'to think'), and not *psyche*. Or does it? Perhaps not even 'noology' goes far enough. **Scientific materialism can, by definition, only rationally (i.e by its mode of reasoning from the *PAM*) allow physical composition.**

Upper, Psychological Sub-divisions of the Ziggurat

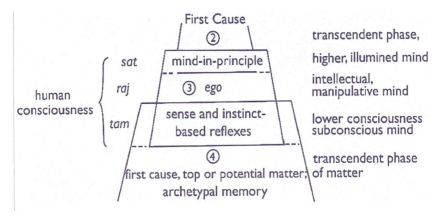

So what does the word 'soul' or 'psyche' mean? Inner essence? Centre? Does it include immaterial consciousness? *Bottom-up*, it is figment of brain's motion, an imagination born of nerves.

Top-down, however, might you not reflect creation as a whole? Is your own constitution, <u>microcosm of the macrocosm</u>, in the image of Chapter 2's three-tiered universe and Chapter 4's Informed Man?

Having earlier registered qualities of *Psyche* (Soul) and *Logos* (First Cause) it is at this point noted that, from a *top-down* point of view, created 'mind-with-body' is seen as a 'soul vessel'. Thus every different pot (or organism) contains the same pure water. In this manner a human may be seen as soul having a physical experience rather than a body erroneously imagining its metaphysical soul. Also, in this view, who masters best his subject is best qualified to treat - top-class and wise soul-doctors (called psychiatrists) are saints.

Clearly, we have two very different views of mind and soul.

Substitution, metaphysical for physical, is, however, scientifically 'unacceptable'. 'Impossible' - whatever future studies might disclose. **'Immaterial' is a word too far.** *So, naturalistic creed decrees, thought's entirely a nervous matter.* ***Thus, naturally, psychology emerges as the study of neurological phenomena.*** Carbon, oxygen and more, from this perspective soul is the activity (incredibly, incomprehensibly complex, mind you) of particles. Consciousness 'emerges' from non-conscious molecules grouped in some special way. Isn't thinking generated by the soft-wired workings of a brain? *Just sling sufficient atoms, in the form of nerves, together - they'll become no less than self-aware!*

Holistically, however, the notion that mind is a physical illusion is the neurological delusion.[28] **To mistake its neural correlate (as, say, registered in brain scans) for experience itself is an error as basic as taking the electronic pulses in a wire for all there is to telephonic conversation. It is a prime, elementary fault, a first category philosophical mistake; it might be termed full-blown, psychological mythology.** ***In this case, the scientific phrase 'neural basis of consciousness' is fundamentally incorrect;*** **and** to identify consciousness as a material illusion is itself, denying the reality of one's own experience, a pernicious - even dangerous - delusion.

One party, it is clear, *believes life is a phantom of the atoms*. Brain *causes* mind. Thought (therefore belief and all the purposive effects of hope and will) is part and parcel of nerve chemistry. Subjectivity is thus objectified! What, therefore, is the *experience* of consciousness? The essentially robotic view of neuroscience holds that nerves *are* consciousness. We just don't yet understand, the faithful purr, how brain's 'emergent properties' can squeeze experience out or how the juice that's 'you' must be exuded from its molecules. A revelation is, however, prophesied. Materialism's scientific certainty decrees that life will be reduced to chemistry and mind experimentally identified as simply due to complicated ionicity! Subjectivity is thus objectified. You are a product of your physiology and so, at root, your genes alone. Life has, hasn't it, to be an electronic after-thought?

Nervous particles and atoms aren't, like atoms anywhere, alive. Therefore, if life is made of them it shouldn't be alive! *Thus the other line suggests, conversely, that brain isn't an originator but a mediator. It is our dashboard as we fly through life.* A filter. A sophisticated interlink between mind, body and the latter's physical world.

If so, it is a *transducer device* that, like any mediation network (e.g. radio), must be sufficiently well-constructed to handle large volumes of two-way traffic. It accepts environmental signals and translates them (↑) 'upward' into mind's experience; and issues orders (↓) 'downward' into body chemistry. As an organ of 'cockpit control' its 'dashboard' accurately connects an immaterial mind to a material body and, thereby, physical conditions. In this view mind and brain, although compounded, are quite separate entities - the former metaphysical and latter physical. Brain chemistry's identified as a design that expedites exchange of information. Your head is thus a medium!

Making no material difference by adding immateriality, the Dialectic simply reconstructs creation on the basis of a 'conscio-material' duality. **In short, perhaps brain neither does nor ever did enjoy a seamless, subjective experience.** *The implications of this seminal idea are so extensive that this whole course is exploring them.*

Consciousness

So now to Consciousness.[29] This is what it's all about. Without it you are nothing. The star of every play is mind; the kingpin of psychology is consciousness. What is the 'thing'?

First and foremost, consciousness unifies; its very nature, both in experience and intellectual endeavour, is to unify. To make overall sense. Isn't your own conscious mind what unifies, as you, the world around. This is life. **Subjective consciousness is the Great Unifier, the Great Connector.**[30]

On the other hand, non-conscious matter is a special case of its subjective absence; *it is pure non-consciousness.* Clouds, streams and solid bodies don't know anything; and, as matter's pure non- consciousness exists, could not a concentrate of immaterial information - pure consciousness - have being too? Elements of wakefulness or knowledge are not physical. So that creation's source turns out to be material oblivion's antipode - *Potential Knowledge, Latent Field of Knowing, Pure Consciousness. Non-conscious oblivion's polar opposite is Total Wakefulness.*

Hierarchical Psychology: Mental Ziggurat

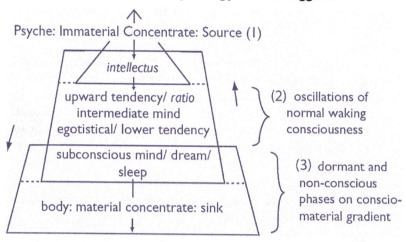

Tier 1 is Source, tier 3 is sink. And, from *top-down* perspective, any attempt to reconcile physical science and psychology must logically start at the top rather than bottom extreme. It must start at Axis, Centre, Source or Cause (Conscious Soul) rather than with any subsequent effect. *Psyche. Soul. Potential Information. Uncreated Consciousness is precedent and therefore primary; the lower, existential range of forms and motions follows secondary.* **What light is to physics Wakefulness is to psychology; in other words, the only absolute measure of consciousness, and therefore psychology, is Transcendent Super-Consciousness.** Therefore of what, you ask, does *total wakefulness* consist; what might be the nature of Pure Consciousness? Could the Experience of such Concentrate be what, mystically, is labelled Love, Pure Love?

Materialism doesn't like this sort of phraseology at all.

tier 3/ tier 2	*Tier 1*
objective/ subjective effects	*Subjective First Cause*
lesser levels of consciousness	*Pure/ Ultra-Consciousness*
do/ know	*Be*
body/ mind	*Soul/ Pre-active Potential*
↓ tier 3	*tier 2 ↑*
passive, programmed object	*programming informant*
non-conscious/ reflex	*subjective/ conscious*
body/ physical activity	*voluntary/ psychological*

It's far too paradoxical. **Nevertheless, the stack above illustrates triplex psychology.** Its three phases are Super/ Ultra-conscious, conscious

47

and sub/ infra-conscious. These 'gears' in the system act like levels in a cosmic grid, a chain of command. This chain applies equally to our microcosmic selves. **It maps, as simple, hierarchical psychology, onto the mental ziggurat above.**

We can now descend from the level of super-state transcendent (*psyche*) briefly to the place of our restricted, human awareness.[31] Next we fall to cover, at the sub-conscious end of mental balance, the conditions of dormant mind, dreaming and deep-sleep; finally we inspect the psychosomatic domain of instinct, personal memory and archetype. The latter account for the psychosomatic connection between mind and its sub-state, non-conscious material body.

These 'bands' or phases of descent are divided slightly further to relate *five* main states of mind with their correlated conditions of brain. Which 'gear' are you in right now?

Five Main States/ Gears of Human Mind in Relation to Brain

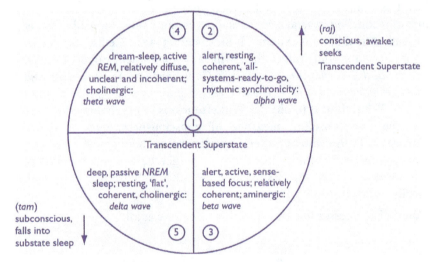

Vectors and Five States/ Gears of Mind by Stack

spectrum of consciousness *(2),(3),(4),(5)*	*Super-Conscious* *(1)*
↓ *downward tendency*	*upward tendency* ↑
passive/ exhausted	*active/ alert*
inertial equilibrium	*dynamic equilibrium*
non-conscious/ subconscious	*conscious*
busy-asleep/ dreaming theta (4)	*busy awake beta (3)*
deep sleep/ coma/ delta (5)	*contemplative/ alpha (2)*
involuntary	*voluntary*

Of these states, as also shown in the previous stack, 1 relates to Super-consciousness, 2 and 3 to waking and 4 and 5 to sub-consciousness.

Libraries have been written concerning the state of psychological normalcy, our waking state (here 2 and 3). Suffice here to note that *ego* is a mask or per-sona. Its dynamic structure, necessary for function in a body, amounts to a 'face' or formful covering over inner, underlying consciousness. It is this 'face', this constraint of ego that a contemplative attempts to 'vaporise'.

We are now in a position to connect metaphysical with physical hierarchy, the latter as it appears in the construction of the most complex object in the universe - a human brain. Claimed to have evolved out of a total lack of logic it can, however, produce a logical and comprehensive understanding of itself. Boot-strap logic here, *par excellence*!

Here is brain drawn simply.[32]

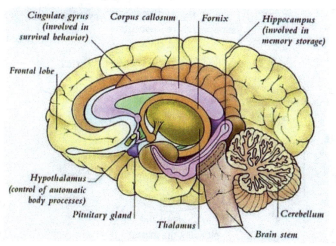

Bottom-up, brain is seen as the generator of thought. Nerve cells, made of non-conscious molecules, create the experience of an illusion known as you.

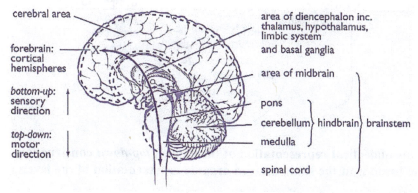

***Top-down*, on the other hand, brain is seen as a mediator, a complex dashboard, a fine linkage factor between physical body (of which it is a part) and metaphysical mind (to which, for the body's lifetime, it is attached). The brain is a non-conscious machine, a mind machine like a computer. Reflex calculation is its task.**

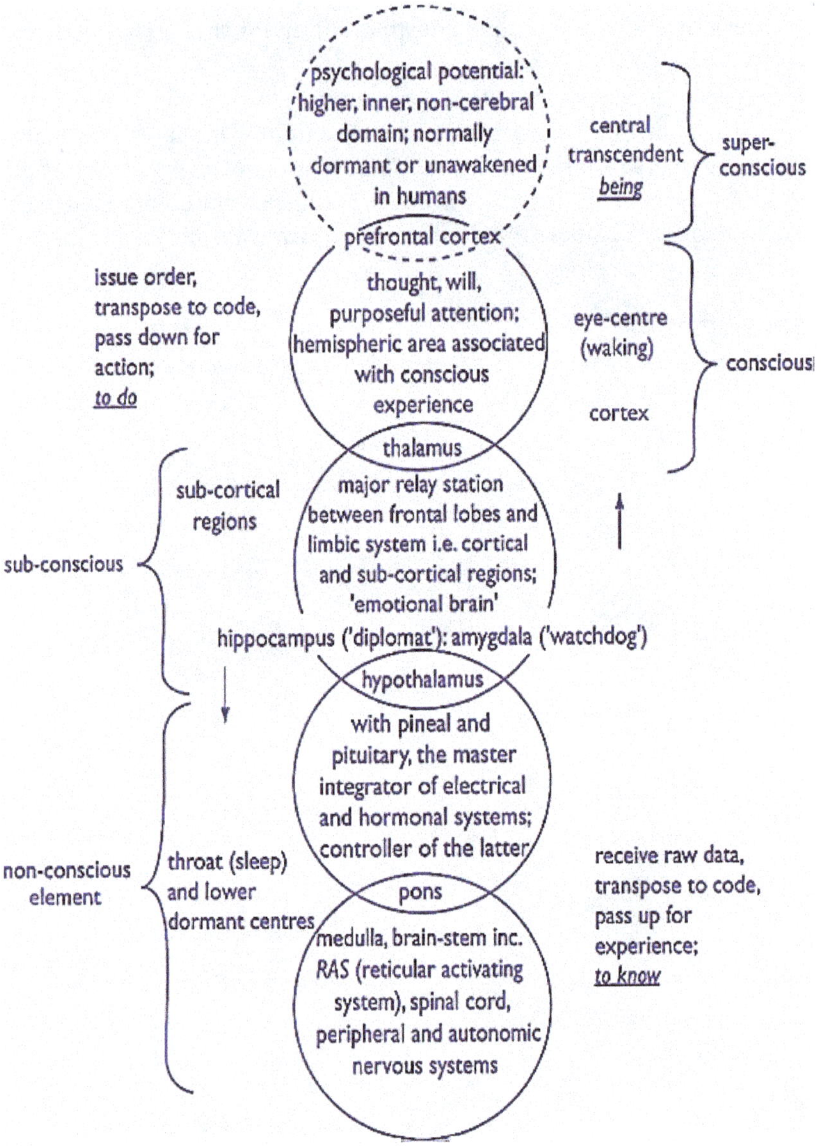
psychological potential:
higher, inner, non-cerebral domain; normally dormant or unawakened in humans
prefrontal cortex
thought, will, purposeful attention; hemispheric area associated with conscious experience
thalamus
major relay station between frontal lobes and limbic system i.e. cortical and sub-cortical regions; 'emotional brain'
hippocampus ('diplomat'): amygdala ('watchdog')
hypothalamus
with pineal and pituitary, the master integrator of electrical and hormonal systems; controller of the latter
pons
medulla, brain-stem inc. RAS (reticular activating system), spinal cord, peripheral and autonomic nervous systems
central transcendent being
super-conscious
eye-centre (waking)
cortex
conscious
issue order, transpose to code, pass down for action; to do
sub-cortical regions
sub-conscious
throat (sleep) and lower dormant centres
non-conscious element
receive raw data, transpose to code, pass up for experience; to know

This dialectical representation of the logical, *top-down* construction of the brain is at the same time an organic representation of the levels

of mind. Are these illustrations linking brain structure to cosmic hierarchy? Take a little time to work out what, *top-down*, is being said.

Was brain thought out? Is there, after all, no logic in the way a brain is built; is its reflection of the hierarchical order of creation just an aggregate of happy, codifying accidents? *It is logically surprising that, for no reason, non-conscious and illogical chaos should have constructed order to a very highly systematic climax in the most complex working system of the cosmos, an information processor whose whole, sole, negentropic business is order - a central nervous system and associated brain.* Did matter, getting far more than it didn't bargain for, perchance 'evolve' the hierarchical control reflected in a brain? Materialistically speaking, it must have. *In reality did it?*

A brain so simple we could understand would be so simple that we couldn't; but, *top-down*, we can enumerate the principles round which its great complexity might gather.[33] Relate this illustration to the three preceding it. Does not the order of its parts confirm a *top-down*, hierarchical construction of our dedicated organ of intelligence?

From conscious mind we now descend to the large subject of *sub-consciousness.*[34]

Subconsciousness

The Sub-conscious, Psychosomatic Sandwich

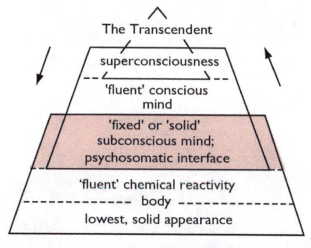

Are you ready for a fall to the diffuse conclusion of psychological entropy, for a subjective drop into the labyrinth of underworld? You know what it is to be mentally as well as physically exhausted. You've often dropped off into mind's flat, dark condition we might call inertial equilibrium. 'Little death' is not the world's end so let us take a snooze cruise; it is time to fall asleep.

Dreaming. For dreamers dreams are real enough; but the experience is untrammeled by either external events or the ability to reason. Waves wash equally on what is in their path; a torch shone randomly around picks out disconnected or illogically connected objects and events. The files are scattered, narrative is blurred.

Deep Sleep. In deep, non-*REM* (or *NREM*) sleep the 'upper', voluntary structures of brain are cut from the loop. A sleeper's movements, including eye movement, are much restricted; sensation is dull or absent. Brainwaves, the overall coordinators of the central nervous system, slow to between 0.6 and 3 hertz. These are so-called delta waves. Maybe deltas drop to zero. Brain death. If, by head injury, stroke, tumour or poison, the sleep/ wake toggles fail or signals cannot reach the forebrain then the patient drops into oblivion. The curtain falls but drama does not start again. Coma is an open tomb, an unpinned shroud or wake-less sleep - though in its stillness deeper grooves of mind (archetypal constructions but also profound personal impressions and rote-learned habits such as language) stay frozen yet intact.

Frozen Time. You sleep but your past does not disappear. You wake and your past has not disappeared. You think you have forgotten, you may even suffer amnesia but untapped memories remain. They are how we freeze time. **Memory is frozen time; it is 'solid-state' mind whose 'chip' symbolically encodes the past.** *And a memory is a thought object that, as such, has no life of its own.* A disc encodes music once recorded 'live'; it's a memory that, when replayed, affects the present and, from this, the future. Thus mental memory is an encoded image; it is a record and, on conscious recollection, brings past to present and may so affect the future.

Lower, Unconscious Sub-divisions of the Ziggurat

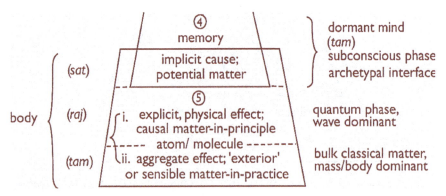

From a top-down perspective, **neither memory nor knowledge are inherently physical.** Mind is a wireless entity. No doubt, correlated nervous circuitry acts as a storage-and-retrieval system that, by association, allows the immaterial library of remembrance its efficient, selective interaction with an innervated body and its circumstances. Thus so-called 'engrams' (are they sited at synapses, inside a nerve cell's body or elsewhere?) may, if they exist, indeed *relate* to physical experience; they may act as a recognition trigger, reference point or body's resonance with an experience. And organs (such as hippocampus and amygdala) certainly seem to log experience in the manner of a record/ playback head; they catch or release a moment that, in fixity, is called a memory. But if such 'storage' or 'playback' button fails the system's compromised. Either records are not made in the first place or the connection becomes impaired or irretrievable. *But the 'disc' of memory itself is metaphysical.*

Memory, the only form of metaphysical information storage, is the shape of infra- or sub-consciousness. *Indeed, it <u>is</u> sub-consciousness.* <u>Subconscious mind consists of coded files that we call memories. It is made of memories.</u> Regarding life-forms memory comprises an organism's library of precondition and conditions, that is, its context for experience. The precondition is its archetype, the basis of its sort. We might call this sort of memory '*typical*' while '*personal*' experience includes both active (created and transmitted) and passive (received) information. Either kind of memory is 'frozen' like still photograph or film. **It may operate like a movie or store programs that, once triggered, can unfold like stories in a sequence to their completion.** Who has ever committed a poem to memory? Any mentally stored plan is such a memory. Programs are, although dynamic, still a frozen form of mind; and they're replete with information. They specify the most efficient means to a well-defined end. **You might argue in this vein that biological structures are codification incarnate; and that the concept they express is an archetypal program**.

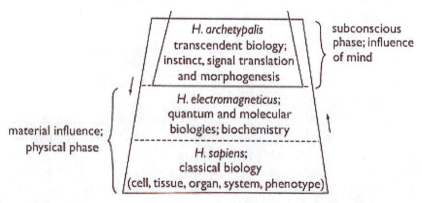

Archetype[35] we see as a **First Cause**. There are First Causes, psychological and physical. <u>**The latter's archetype (potential matter) will, along with quantum matter-in-principle and matter-in-practice,**</u> (see Glossary and Chapter 4: lower sub-divisions of the ziggurat) <u>**lead us to a triplex view of the bodies of physics and biology.**</u>

This diagram illustrates lower sub-divisions of the ziggurat in terms of internal hierarchy and the place of typical mnemone or *H. archetypalis* in biology (see Glossary: mnemone). *H. electromagneticus* and *sapiens* are dealt with overleaf and in Chapter 5.

Archetype (see also Glossary: *tanmatra*) is, like any idea, the informative potential for its own later expression. The word means basic plan, informative element or conceptual template. Its type is, according to holistic logic, held in subconscious mind; archetypal memories compose the subconscious element of universal mind. Thus it consists of pattern in principle; it is the instrument of fundamental 'note' or primordial shape, the causative information in nature or 'law of form'. It is nature's *bauplan* or blueprint. **Called potential matter its fixed files are seen as hard a metaphysical reality as, say, particles are physical realities**.

Of modern psychologists Carl Jung had perceptive sympathy with the idea of archetype. Although he did not cast his net wide enough to include all organisms and the whole inert part of cosmos he nevertheless suggested that human archetype involves elements of the 'collective unconscious' or 'universal psychic structures'. Such 'psychophysical patterns' (in terms of typical mnemone invariably fixed like an IC chip or a broadcasting channel) are given flexible individual and cultural expressions by, in Jung's restricted example, human consciousness.

Furthermore Jung uses the analogy of light's visible spectrum wherein normal human consciousness is yellow with unconsciousness at both red and blue ends: Natural Dialectic's conscio-material gradient likens the visible spectrum to normal mind with ultra-violet scaling super-consciousness and subconscious infra-red grading down to extra-low frequencies of non-conscious matter. Thus Jung also sees the archetype as a bridge from mind to matter. Finally, crucially, just as local objects and events are details distinct from the general principles that govern them, he differentiates between specific, local, conscious imaginations (or 'motifs') and the 'unconscious' generality of archetype from which they derive.

Not only Carl Jung but founding fathers of neurophysiology addressed the mind-matter (psychosomatic) issue in which immaterial archetype plays, dialectically, such a pivotal role.

"What greater inherent improbability", wrote Sir Charles Sherrington, **"than that our being should rest on *two* fundamental elements than on one alone?"**

In this case, how does mind interact, primarily at least through brain, with body? What might constitute the nature of an interface (*PSI, see* Glossary and One World: lecture 4 *passim*)? **We turn to the bio-classification of psychosomatic linkage.**[36]

Three diagrams help illustrate a suggested architecture of the subconscious[37] for both conscious and unconscious organisms.

Grades of Human Bioclassification

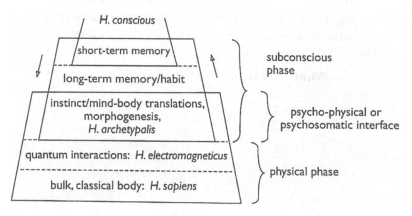

Firstly, your own graded, dialectical bio-classification. This ziggurat represents both physical and metaphysical parts. Its typical memory (*H. archetypalis*) is a read-only cache laid down in universal mind. Not so personal short and long-term memory. In this case incoming (sense) data impacts cerebral nerves and creates localised nervous patterns linked, by **signal translation**, to conscious experience; and, in the other direction, to outgoing cerebral patterns of motor response. Any read-write data file needs be tagged for future retrieval and the suggested label is created, automatically, by the nervous configurations created by, and so associated with, its cause. Whether this localised 'address key' or 'pin-code' is logged with a central reference library (at, say, the amygdala and/ or hippocampus) is unclear. However, if some key configuration is triggered by a later, similar train of thought, emotion or sensation, the memory (that is, the image of the original experience) will be automatically 'alerted' and perhaps fully retrieved. In the reverse direction, a memory retrieved into conscious mind can also trigger responses according, by computation, to the cerebral locations of the nervous configurations that originally responded to a given perception. In other words, a memory retrieved to consciousness can back-track to trigger the physical associations that accompanied its creation. At these key localities the instrument of resonant association (which we'll deal with shortly while discussing 'cymatics') would reiterate the same or similar psychosomatic algorithm as first laid down. Thus, while the bank of personal files in metaphysical memory would remain intact, physical degeneration of nerves could impair the process of tagging by address keys and subsequent physical responses by way of such faulty keys. Now, what do you think about the mechanism of memorisation? It's time for creative (as opposed to critical) ideas of your own.

55

As every cell contains chemical genetic information, so it is suggested (One World: lecture 4) that each also acts, regarding the various electronic configurations of its molecules, as a linkage to its typical mnemone, that is, its metaphysical broadcast. In other words, as well as by 'holographic' *DNA*, each cell is informed at its psycho-physical level by subconscious, archetypal memory. Every cell of every organism involves mind. Typical mnemone is a body's *'metaphysical DNA'*.

Wireless Man

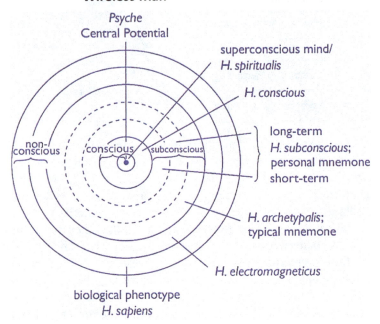

Secondly, this same bio-classification of hierarchical, inward levels of human being can be illustrated using the model of concentric rings. Outermost material and inner immaterial body-shells are shown as the rings of Wireless Man. Its typical mnemone (*H. archetypalis* or 'memory man') is, in effect, a fixed psychological/ metaphysical structure channelling the program of human **instinct**. Positioned at the mind's side of the ring representing psychosomatic interface, its physical correspondent is *H. electromagneticus*. Peripheral *H. sapiens* is, in a radio world, the sole 'hard-wired' part.

Thirdly, as shown in Two-Way Psychosomatic Linkage, typical mnemone is composed of three main sub-routines; or, if a protocol is a standard procedure for regulating the transmission of data between two end-points, three linked protocols. Together morphogene, instinct and personal data storage comprise an individual's archetype. If, as in the case of some animals, plants, fungi and bacteria., there is little or no conscious component, then the influence of typical mnemone is reduced to instinctive and morphogenetic components (Glossary: morphogene). **More details of suggested systems architecture can be found in Appendix 3 and elsewhere.**[38]

Signal Translation: Two-Way Psychosomatic Linkage

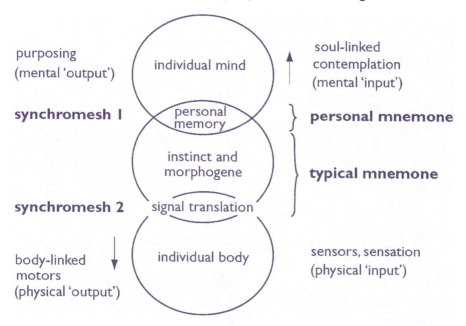

In summary, the following paragraph suggests how the medium or interface of archetypal mind is linked with matter. *As noted in the previous diagrams, mind (at its most gross, subconscious level) conjoins with matter (at subtlest, least-massive/ almost-immaterial level).* *Elusive quantum probabilities pinned-down substantiate, it seems, certain processes; photon, electron and their effect on quark (protonic) or atomic position precede, in the sense of govern, molecular and bulk reactivities; and, where electrodynamics describes the effect of moving electric charges and their interaction with electric and magnetic fields, biological electrodynamics precedes all bio-molecular considerations.* **Every biological process is electrical; and the flow of endogenous currents is the primary and not secondary feature of physical life. Not only biochemistry but quantum biochemistry heave to the fore. Natural Dialectic lifts perspective from molecular to a vibratory, field perspective.** *It is thus suggested* (One World: lecture 4) *that, at electrical and wireless levels, patterns of subconscious mind meet and influence matter; archetypal information is relayed to chemistry by polar charge and electromagnetic light.*

In the previous diagrams this subtle, wireless relay (in the human case *H. electromagneticus*[39]) is identified as the physical side of message-exchange; and *H. archetypalis*, with its routines, the metaphysical calibrator.

There is no doubt, the universe is in vibration. The cosmic transmission of information and energy is, at root and as spectroscopy confirms, vibratory. **Ordered oscillation is called harmony; harmony is the grammar of music and music is a universal language**. The theory of music is implicit in any recital. Could it transpire that explicit order of a cosmic recital is the product of implicit notation? Could its performance represent cooperative forces, specific 'notes' called particles and thereby, all in all, harmonic code?

Here we note the work of, among others, Ernst Chladni (above) and Hans Jenny. **Their study, now called cymatics, involves the use of wave frequencies to precisely control the production of shapes in water, air and other media.** It is significant that Chladni's figures often imitate familiar organic patterns that we see in nature and, especially, biological structures.[40]

 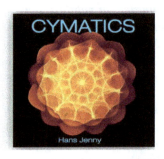

Pattern clearly relates to frequency of cycle. And resonance, whose orderly aspect is characterised by an analogy with music, is the tendency of a body or system to oscillate with larger amplitude when disturbed by the same frequencies as its own natural ones. *Cymatics therefore intimately involves the vibratory transfer of energy.* Such transfer is an integral part of all vibratory systems wherein waves interfere with/ destroy or cohere/ amplify each other. Energy is amplified and transferred by resonance and attunement. Common examples of electromagnetic resonance include tuning a transmitter/ receiver and photo-electric initiation of the photosynthetic process, that is, the first step in life's chemistry.

For quantum physics matter is certain vibratory frequencies of energy. From a dialectical point of view, it is simply stresses, strains or tensions in the medium of their absence, that is, in the extent of space. **There is, however, nothing random in a highly orderly creation derived from first acoustic principles. Oscillation between polarities, cycles, vibratory rhythms, <u>the interrelationship of waveforms and complementary resonance are at the heart of Natural Dialectic</u>.**

And it is suggested that a key phrase in the suggested explanation for the wireless, psychosomatic transfer of information between subconscious structures of the mind and the physical plane is, at root, '<u>resonant association</u>'. *The modus operandi of psychosomatic broadcast is therefore, in a word, attunement. Resonance.* It involves entrainment between typical mnemone and *H. electromagneticus* or, if you like, transduction between recorded information (memory) and physical energy.

Along with instinct and signal translation the *third* component of a typical mnemone is its **morphogene** (see Glossary). Of course, materialism ridicules the idea of natural but also metaphysical informative structure, a 'cloud' or mnemonic database called archetype. You might well, if Hippocratic *vis medicatrix naturae* is other than chemical, deny its clinical influence. Because what cannot be physically explored does not physiologically exist.

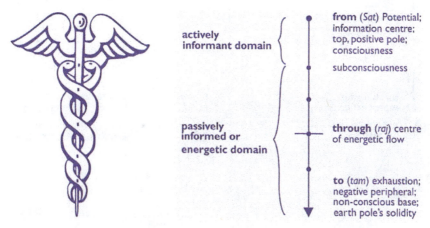

actively informant domain

from (*Sat*) Potential; information centre; top, positive pole; consciousness

subconsciousness

passively informed or energetic domain

through (*raj*) centre of energetic flow

to (*tam*) exhaustion; negative peripheral; non-conscious base; earth pole's solidity

No doubt, *DNA*, hormones and other codified biochemicals are intimately linked with growth, development and morphogenesis but are

59

they all there is? And do we suppose these microsystems or a cell's or body's perfect health evolved through grades of imperfection, that is, millions of cacophonuc, non-integrated, sicker stages? Or, turning that 'progressive' notion on its head, do you prefer that archetypal vigour radiates original, dynamically fine-tuned health for every kind? *If your body's health is indeed the result of interaction between physical and archetypal (sub-conscious) patterns, it did not evolve by accident.* The tendency of every system is to bounce back to its archetypal norm, to heal itself.

This metaphysical form is the very basis of homeostasis, physical life's overriding principle. Health is co-orderly; dynamic equilibrium is the norm. Wounds heal, infections are fought and cell debris cleared. Indeed, the central nervous and autonomic, endocrine and immune systems are integrated in such a holistic way that experts sometimes use the phrase 'psycho-neuro-immunological system' to describe their cohesion. And mind's attitudes and purposes substantially affect the body. The channels of 'psycho-somatic interaction' are precisely what this chapter is describing - unless you still believe that mind is brain and psychological equates, essentially, with chemical.

Informed Man

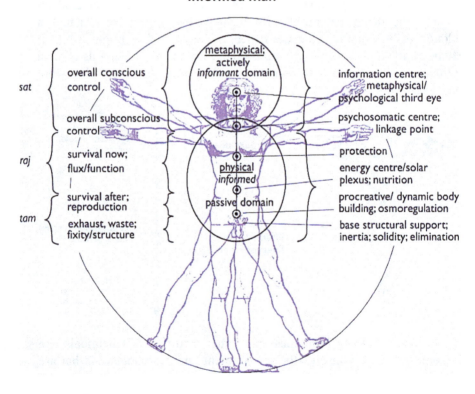

sat { overall conscious control

overall subconscious control

raj { survival now; flux/function

survival after; reproduction

tam { exhaust, waste; fixity/structure

metaphysical; actively *informant* domain

physical *informed*

passive domain

information centre; metaphysical/ psychological third eye

psychosomatic centre; linkage point

protection

energy centre/solar plexus; nutrition

procreative/ dynamic body building; osmoregulation

base structural support; inertia; solidity; elimination

Informed Man: *see* also Glossary *prakriti, prana, chakra.*

.Informant archetype is 'memory man' and, as such, shapes 'also '.[41] The above-pictured wingèd staff, ancient symbol of Hermes, remains to this day the totem of many doctors. From brain to base of spine it represents the human extent. In the way of Natural Dialectic, its gradient reflects the linkage between informative and energetic domains of humankind. The microcosmic slope runs top to tail - from informative head section through an energetic centre (heart, lungs, digestive tract) to base systems of exhaust; from informant to informed by way of double helix it illustrates the logic of embodiment.

Why should a fundamentally rational system irrationally 'self-organise'? Your own human microcosm underlines the utter feebleness of evolutionary explanation. **If materialism's rational you could argue that it spotlights how irrational, backing chance as its creator, this 'counter-intuitive' species of rationality has become.**

Isn't it, on the holistic hand, inconceivable that such a logical, integrated, self-consistent embodiment as yours, constructed with highly specific complexity in accordance with grades of the conscio-material gradient of creation itself, occurred by chance? **If reason wins, whose archetypal program is worked out in every bio-form, then chance and time's grand evolutionary theory crumble back to their home ground - they bite their progenitor, the dust.[42]**

Questions 3

1. What three items compose the hierarchical triplex of psychology?

2. Arrange them as the triplex member of a 'stack' (Appendix 1 might help here).

3. What is the nature of soul to a materialist?

4. Where and what is the nature of trans-existential soul - if it can be imagined?

5. Which of the triplex does 'scientific' psychology treat?

6. In what form does materialism claim non-consciousness can become conscious?

7.   What is 'the neurological delusion'?

8.   Does the brain know anything?

9.   Where is the seat of consciousness in the body?

10.   How would you define 'higher mind'?

11.   What is the difference between 'horizontal' and 'vertical' causation.

12.   An idea 'vertically' precedes a physical action. Has this action (or rearrangement of physical material) the capability of occurrence without mind?

13.   Can mindless activity occur?

14.   What is potential?

15.   How is 'archetype' related to potential?

16.   How many cosmic archetypes/ first causes are there?

17.   What are the five states of human mind?

18.   What is *ego*?

19.   What do we call a frozen piece of mind?

20.   What is subconscious mind?

21.   How is brain constructed to reflect the order of an act of creation (and, in the opposite direction, perception)?

22.   Is a brain, developed from a single codified cell, likely to have obtained its coherent, efficient and purposeful architecture by aimless aggregation gradually over time?

23.   If mind is immaterial and body material how might they influence each other?

24.   Of which shells (or sheaths or bodies) do you, wireless man or woman, consist?

25.   How many of these shells are physical?

26.   What is the junction called 'synchromesh 1'?

27.   What is the junction called 'synchromesh 2'?

28.   What is a mnemone?

29. What are the two divisions of memory?

30. Memories are forms in mind. What are the three cooperative forms of device employed by a typical mnemone (or archetype of an innervated organism to support the operation of a conscious mind?

31. Where is the typical mnemone, a feature of subconscious mind, found?

32. What is cymatics?

33. If mind is not matter, then psychosomatic interactions must be occurring at the rate of parallel transactions in a non-stop super-computer. What is the suggested explanation for direct psychosomatic transaction?

34. Suggest how the mechanism for personal memorisation might work.

35. Is the arrangement of your body parts logical or not?

Chapter 4: Physics

Friend Lao Tse now with Niels Bohr (the quantum theorist who, having showed that the electron orbits of an atom explained its chemical properties, neatly linked physics and chemistry) remind us that Dialectical foundations have stood the test of time. Bohr incorporated the dualistic, dialectical *yin-yang* design into his family crest!

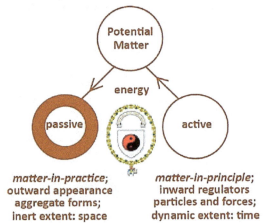

Other modern seekers after physical truth include:

Newton, Einstein and Planck. Now note three points:

1. Humans are information-cravers. We want to know more and more about our phenomenal surroundings and the working of our own minds. We want, moreover, to find patterns by which to order our knowledge to reflect as closely as possible the way in which the world works. Check the blue boxes of Chapter 1.

> **Physics, however, is restricted to answering questions purely with respect to and in terms of the non-conscious, reflex, physical world in which we find ourselves.**

2. Physicists are aware that, despite great advances, there is much we do not know. Although in 1900 it was thought that physics was, essentially, complete, 2000 radically disagrees. Apparently we have only studied 5% of the universe, luminous matter. Here, not mention how the universe began, is a sample of sub-areas of unresolved difficulty in physics:

Dark matter	Dark energy (lambda force)
Nuclear physics	'Flat' space
One or parallel universes	Wave function collapse
Grand Unified Theory	Quantum gravity
Magnetic monopoles	String Theory
Sonoluminescence	Superconductivity
Dimensionless constants	Gravity and gravitons
Holonomic brain theory	Inside a black hole

3. Mankind may eventually completely understand the cosmos it inhabits but, given the above, don't expect full answers to every question in an hour or two! This chapter fits physics, as far as we understand it, into the frame of ND (Natural Dialectic). It gives fresh context and perspective. This especially applies because material science cannot reach actively informative, metaphysical but still natural elements; and it cannot include this immaterial 'missing factor' in its measurements. Such factor applies to matters psychological/ metaphysical but matters not one jot for accurate physical appreciation of cosmos. However, because abductive (best-guess) logic and interpretation are required to deal with historical events - especially original and consequent order regarding cosmos and biological (mind-matter) forms - material explanations alone may turn out incomplete. **Perhaps, indeed, human mind is not built to understand the beginnings**

of things but, being ever-curious, we'll work towards our best shot, that is, our holy grails.

What, dialectically, are holy grails about? They are about achieving, at the heart of myriad, detailed complexity, simplicity and union. They are of two kinds - material and immaterial.

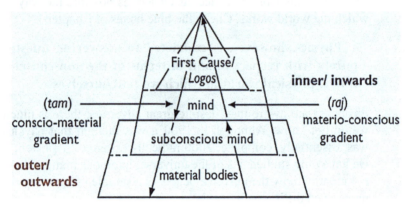

Do you remember this key illustration from Chapter 2? Scientists of physical phenomena and scientists of the soul travel in *opposite* directions. In spite of this both seek a Source and, as such, First Cause; but their perspective differs and their goals, immaterial and material, are actually antipodal. **This does not mean both are not, in their way, right or that a man cannot be both scientist and a mystic.**

Inward, immaterial concentration here.

The latter repeatedly exercise their contemplative faculty at its inward focus; seeking 'inside information' they travel *up* a materio-conscious gradient away from physic towards pure metaphysic, towards the source of mind. Such grail is called enlightenment, that is, unification in First Cause. This Cause has, as we saw in Chapter 2, many names. The *Logos* is the *Tao* is The Informant. **It the Most Natural Heart of Nature whereas physical archetype is the natural heart of non-conscious physic. In as much as**

Natural Dialectic provides a vehicle within which the root causes and grails of physics, psychology and biology may be coherently co-understood, it may be called a Cosmic Unification Theory. Including *both* mind and matter it is, in effect, the *CUT* of *GUTs*.[43]

Material exploration far and near.

Scientists concentrate, as we mostly do, *down* the conscio-material gradient; seeking 'outside information' they follow out through the senses and their technological extensions to learn the details of creation's unconscious shell, the material universe - including our own physical bodies. They seek, in a quest to find the **Holy Grail of Physics**, the implicit cause of explicit, physical phenomena.

Holy Grail of Physics

subsequent hierarchy	Holy Grail
issue/ action	Source/ Priority
physic	Metaphysic
phenomena	Causal Noumenon
classical/ quantum	Archetypal
↓ tam	raj ↑
outward finality	inner support
exhaustion	action
aggregate/ precipitate	influential action
large-scale body	microscopic components
matter-in-practice	matter-in-principle

This quest, driving to the simplest basics, is called unification. **Such unification (by *GUT* or *TOE* - a 'theory of everything'), is the goal with its subsequent hierarchical organisation; but the maths of microscopic** matter-in-principle (quantum theory) and large-scale matter-in-practice (relativity) do not fit together. A joiner might be **quantum gravity** and theoretical attempts to create a union are taking place. Nor are dark matter and lambda force (anti-gravity) much in the picture. It seems an ephemeral Higgs boson may, at very high energies, exist but it is more hoped for than clear how this tiny beastie's field produces that most basic of physical quantities - mass - and, if not, how did mass appear? Nor, of course, has any immaterial factor ever been considered. Universal mind is not considered. **Now, however, in the attempt to integrate mind with matter, let's investigate this stack of physical holy grail.** Its three parts (metaphysic, matter-in-principle and matter-in-practice) are interesting because they introduce the notion of a <u>Triplex Physics</u>, that is, three stages of physics according to the cosmic fundamentals (*sat*) informative potential, (*raj*) action-in-principle and (*tam*) end-product, action in finished practice. **In other words, they prompt towards a Theory of Physic and Metaphysic.**

Primary (Metaphysical) and Secondary (Physical) First Causes

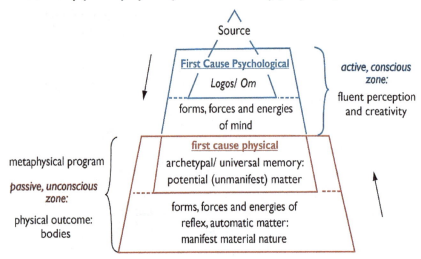

Recall, from Chapter 2, a stack and cones that illustrate the place of Primary and Secondary First Causes; and, from Lecture 3, The Upper Conscious and Lower, Unconscious Sub-divisions of the Cosmic Ziggurat. Do you also recall that psychological first cause is conscious but physical first cause is, although metaphysical, like all the physical universe, unconscious? Such secondary first cause governs physical cosmos. The 'father' of physics, it generates precise quantum forms. As thought is father to the deed or plan is prior to ordered action, so 'patriarchal' archetype precedes material phenomena. **This archetypal grail is called potential energy or, according to Einstein's equation, potential matter.**

In consequence, we turn to deal with **triplex physics** as regards its fundamental levels. We start with *archetype - potential energy*. There follow active matter-in-principle[44], the staple of quantum physics; and passive matter-in-practice, the bulk aggregates we call our sensible universe.

(Sat) Potential Energy

The models and stacks that follow in this chapter are *generalisations* illustrating, from a Dialectical perspective, the order of major physical instances. They drop from the pre-physical first cause, archetype, through the phases of cosmos. The phases are triplex, even in the lowest, familiar subdivisions of the bulk or classical phase of matter-in-practice - gas, liquid and solid.

tam/ raj	*Sat*
physical matter	*Potential Matter*
action	*Physical Absence/ Void*
program played	*Archetypal Program*
lower syntactic level	*Upper Syntactic Level*
↓ *tam*	*raj* ↑
matter-in-practice	*matter-in-principle*
gross/ extrinsic constraint	*subtle/ intrinsic motions*
data item/ physical form	*data item/ physical force*
classical bulk	*quantum flux*

Potential precedes possible action. It is a prerequisite or precondition for results; *but Natural Dialectic does not think of physical potential in the way that physics does (e.g. potential gravitational or electrical energy).* **By now we understand potential energy/ matter is a metaphysical informant.** It is super-matter whose synonym is archetype. **Archetype is, like any idea, the informative potential for its own later expression.** In the sense of *precursor* to expression of a physical reality archetype may be seen as a First Cause. **The word means basic plan or conceptual template.** Like instinct it abides, according to holistic logic, as a *form* or *file* of memory in dormant, universal mind. It serves as physic's data bank.

Thus it consists of pattern-in-principle; it is the instrument of fundamental 'note' or primordial shape, the causative information in nature or 'law of form'. It is nature's *bauplan* or blueprint. **Called potential matter archetypal files are seen as hard a metaphysical reality as, say, particles are physical realities**.

In short, think of archetypes as comprising the *potential* phase of both psychological and physical triplex. **Archetypal source is therefore an important feature of holistic cosmos, the level whence things develop**

orderly. Do you remember, from Chapter 2, thinking of it as a 'seed' which precedes the origin of physicality? In this case you might expect its metaphysical operation to follow different rules from those in physical space and time. And a seed is related to every part of the body developed from it. Could inmost archetypal seed exist, like a kind of holographic '*DNA*' within the body of our starry universe, apparently nowhere - being metaphysical - but actually everywhere and every time at once? If you think 'archetypal first cause physical' is a 'cop-out' in explaining how cosmogony proceeds then I suppose a systems analyst must think conceptual plans for any working mechanism or his working chips are 'cop-outs' too. However, philosophical objection to a program's purpose in no way mitigates the impact of its natural possibility and, if an immaterial element of information exists, it's natural fact.

With or without such cosmic program, we can reasonably claim invariance under transformation. Underneath the world's commotion character of basic rules and parts remains the same. Contexts change but automated rules of play do not. *And without conserved invariance you can't obtain the balance that equations need.* **Energy's conserved**; **and so is symmetry - at any time in any space from any angle, physicists agree, laws of physics stay the same.** Does physics see such changeless cosmos as a self-consistent, fine-tuned set of principles or are its invariant patterns of behaviour basically informed by chance?[45]

Triplex Physics: Cosmos not Chaos

subsequent orders	*Informative Archetype (1)*
effects	*First Cause*
inertial/ active phases	*Potential Phase*
local motions/ variations	*Ideal Field*
game	*Principles/ Rules*
manifest phenomena	*Unseen Noumenon*
↓ passive phase (3)	*active phase ↑ (2)*
contractive gravity	*liberating levity*
aggregates	*quantum phenomena*
apparent randomness	*internal order*
chaos	*cosmos*

Rules are invisible but they precede a game Rules are information, information is potential for behavioural patterns. Regulation thus precedes and guides the way a game is played spontaneously.

In this case we say that in the beginning was *NOT* chaos. **Modern physics shows that mankind dwells in a finely-tuned universe. To be**

70

precisely fit for life it must have started in a way most orderly, specific, **specially defined.** *Fine-tuned by chance? If low probability together with specific definition indicate design, no chance!* **A universe fine-tuned regarding many parts might be construed as one of specific, irreducible complexity; and, if lacking any part it failed to work, of minimal functionality.**

No objects bear a number yet you number them and count. Numbers, symbols and mathematics aren't a physical but changeless, metaphysical reality. *Maths, while we're on the subject, is a real form of metaphysic.*[46] Pythagoras, at least, believed that 'All is Numbers'. Could essential, physically-independent numbers really govern physical complexity? Einstein, Planck and Eddington believed that, once you dropped upon their key, you could *deduce* (*top-down*) the reasons for all natural laws; you might unlock the codes whose inmost mysteries reveal just how a stable universe is sparked; a feat of mind *par excellence*!

For others, including Roger Penrose, the Platonic world of mathematical forms is also real. He writes, '*There is a very remarkable depth, subtlety and mathematical fruitfulness in the concepts that lie latent within physical processes.*'

Could such immaterials help describe mind-matter frontiers whose formations are called archetypes? **Archetypes are also metaphysical and have no being but in mind. Maybe 'natural mathematics' is describing real, archetypal files.** *Perhaps archetypes compose the link between the corners of what Roger Penrose has referred to as a **mysterious triangle of physics, maths and mind.*** At any rate, their logic's fixed. Fixed forms of mind are memories; thus they are memories in universal mind.

Could *you* plan a cosmos better? Lucid physicists agree it looks 'as if' designed. **When it comes to astrophysics and cosmology stunning ingenuity seems to have coordinated chemistry and physics' natural laws, not least when it comes to bio-friendliness.**

No doubt, in any case, we're here because the universe is as it is. That's no surprise. How came it so? Was its initial condition chance or not? The universal body is sharply defined by a precise set of over thirty interdependent settings. ***Their values combine to generate a universal pin-code that was either preordained or at least intrinsic in primordial projection. Indeed, the dials are set for the sun, earth, you and me to an accuracy computed by Oxford mathematician Roger Penrose at 10^{10} to power 123!*[47] If true, that cuts chance completely out. Erasure of coincidence.** The probability of your bullet hitting a nail-head at the other end of cosmos is vastly greater. Mathematicians consider odds longer than 'only' 10^{50} against to be zero. That is to say, there is statistically no chance whatsoever that cosmos and its dependent life are accidental. **<u>The Penrose</u>**

computation, if valid, indicates that odds against the observed, law-abiding universe appearing by chance are stupendously astronomical; and the facts appear to support his calculation. The consequence of such statistical annihilation is to annihilate theories of origin by chance.

If chance is mathematically intolerable then whence? Where is the projector; what's the transcendent nature of projection's source? Natural Dialectic's 'holographic' edge is everywhere; it's 'super-posed' on physics, omnipresent but invisible because it's metaphysical. **It is the place where theme is turned to individual instance, principle is practised and where physic with its metaphysic meets.** Space,[48] time[49] and local things are *within* universal mind. **Mind's archetypes project our world; archetypal memories, potential matter, are the essence of our physic; they inform, unchangingly, material being. Immaterial information holds the world, physical and biological, together. It is by archetype conserved.**

An archetype is a form of principle. Principles are concentrated information and, therefore, metaphysical power. They describe a source of order or a guiding force for local instances. As rules of a game are invisible on the pitch, nevertheless they substantiate its practice; *Principle is universal, expressions of its practice individual*. In the cosmic case metaphysical archetype governs the character of physical expressions; it runs all players in the mighty, automatic game.

Archetype (potential energy/ matter) can also be thought of as letters in a universal alphabet, as bits, bytes and routines of a computer program, an alphabetic world. Particles and forces form, as physics well observes, the basic letters, punctuation and the grammar of our cosmo-logic's script; their local, contextual combinations form the words, sentences and stories chemistry and physics well observe. Events may, therefore, *seem* not coded; they may *appear* at random but in fact conform to inner, fixed behaviour. Automatic reflex is the product of a cosmic code. Its program is instant everywhere but the 'holographic' generality is locally, specifically expressed. If, however, our cosmos is codified then it has purpose and, as archetypal metaphysic indicates, we need to look beyond non-conscious matter for its 'systems engineer'. Thence speaks the world but does its rationale embody any meaning?

Music, like our speech, is made with vibrant air. Since vibrations, wavelengths or frequencies correlate with forces and particles another natural way to view the world is notes whose interwoven harmonies compose the cosmic opera. These principals can be conceived as carriers of archetypal code that activate the universal stage.

72

'*Magna opera Domini exquisita in omnes voluntatis eius*' is inscribed on the portal of the Cavendish Laboratories here in Cambridge. '*Great are the works of the Lord, carefully studied by all who so desire*' is not a maxim that has deterred a galaxy of scientific luminaries making profound discoveries.

So is it possible that archetypal forms are in vibration and thus, given the cohort of archetypal principles, cosmos is harmonically organised? *Can creation be thought of as an energetic vibratory scale in which its various bands are 'registers' or 'octaves'?* Such structure would allow that sound at a lower, slower, deeper level (say, physical) is played within the same overall composition as higher, lighter psychological forms. Such logic views the archetypal root of cosmos as musical. **Its overall program, or score, could amount to universal opera.**

Such opera is, of course, a reflection and a resonance of its transcendent order. In dialectical terms its Composer is Ideal and its Creator, at the Centre of Nature, First Cause. Song will, as every musician since Orpheus has known, pull you straight to the heart of things, to the centre of life. This is not a mathematical perspective but is it why Sir Brian Pippard, physicist and a Cavendish Professor at Cambridge University, wrote, "A physicist who rejects the testimony of saints and mystics is no better than a tone-deaf man deriding the power of music"?

(Raj) Active Energy

If first cause physical is metaphysical then what is archetype's first physical expression? *Quantum fields* are physics' prior informant - the subject's blank but basic explanation of the universe. These ubiquitous fields are latent voids, potential from which particles and worlds arise. Fields are primary, particles secondary. Forms from formlessness appear. Tensions in space-time, ripples in this holographic ocean, specific modulations like the letters of an alphabet build everything including you.

Thus, in Natural Dialectic's spectral view, exterior appearance stems from subtle inner bands; creation issues, layered, from within. Now, therefore, moving out from metaphysic physics starts. We enter the kinetic phase of active energy, the world of quantum physics we call **matter-in-principle**. Such principle involves, it seems, five kinds of interactive field (including Higgs) and four of particle (electron, neutrino, and two kinds of quark). Such a holy grail, such pinnacle of reductionism yields a long and difficult equation from which everything should be derived - nearly. Not even nearly since it excludes 95% of cosmos (dark energies and matters), inflation and the immaterialities of mind, intelligence, subjective consciousness and therefore life itself!

Our world is not constructed from hard substances but, like music, out of structured energies. Yet this diaphany, claim physicists, is more substantial, subtle and more real than gross effects, a universe of bodies that we sensibly proclaim our truth!

However, from a *bottom-up* perspective, energy and its material agglutinations make up *everything*. Physic is the root from which illusions of the mind and soul have sprung. **Is energy eternal or has it any cause?**

Eternal matter? There was a theory of steady-state.[50] This has been rejected. Divine fallacy (a theoretical god-of-the-gaps) is one thing but instead materialism recommends its own alternative, unholy artifice. It divines an unplumbed, endless stop-gap called a multiverse whence some titanic clash sparked our own universe; pure speculation is invoked to keep materialism on the road. Still, energy is something and, if not uncaused and eternal, must in whatever form have had a start.

Big Bang: A Miraculous Projection

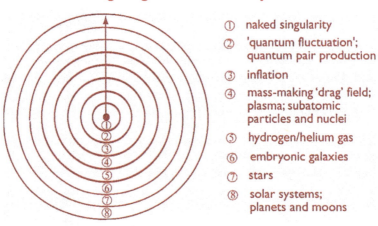

① naked singularity

② 'quantum fluctuation'; quantum pair production

③ inflation

④ mass-making 'drag' field; plasma; subatomic particles and nuclei

⑤ hydrogen/helium gas

⑥ embryonic galaxies

⑦ stars

⑧ solar systems; planets and moons

Therefore, devise one. Current science lacks a reason for emergence into physicality. Its cosmogony tends towards the notion that for no reason, nowhere, absolutely nothing suddenly began 'exploding'. *In other words it relates the phenomenal start of a horizontal, age-related chain of cause and effects conceived of as a 'big bang'.*[51]

Big bang is, in this horizontal sense, a magnificent story of creation with a maths-and-physics-only slant. Bang! Boom! Simplicity itself. **It might be, as we observe from 'inside' physicality, our start-up's smoking gun.** Therefore what headache, even to a physicist, could such a grand yet simple vision of creation present? Sadly, pass the paracetamol. Of nothing-physical, you ain't seen nothing yet.

Who mentions this projection's precondition? *Top-down*, unseen but fecund void, called archetype, is where causation metaphysically, that is, pre-physically begins. *As flower from seed so, in holistic view, exterior expression stems from subtle, inner code; creation is developed, stage-wise, from within.* Nested Centre develops, passing different dimensions of mind and matter, out to its physical periphery - our starry universe..

	grade	time
sat	pre-physical latency	-
raj	quantum micro-level	Planck/ quantum era of high energy and quick time
tam	classical macro-level	era of low energy/ bulk aggregates and slow time

Cosmos not chaos sets initial conditions up. <u>Archetypal files are called potential matter; to reiterate, they are seen as hard a metaphysical reality as, say, particles are physical realities</u>. Plan, principle, directive. Such potential comprises program(s) naturally stored in cosmic memory - simple in terms of inanimate physical 'law' (of particles and forces), complex in terms of animate structure/ function/ behaviour.

Both diagram and stack dialectically express emergent levels. They help to show how archetypal 'nucleus' informs the way that energy precipitates, how certain energetic possibilities are locked into particularities of form and, by thermodynamic generality, material-expression-in-principle (the quantum substrate) and in practice (aggregates in three main states of gas, liquid and solid) materialise as the non-conscious phases of creation's shell.

Grades of the stack rephrase the way 'big bang' is thought to have expanded, cooled and thus precipitated out the world. They may conform to a *top-down* projection of the 'vertical', informative kind They can equally be construed as products of materially energetic, 'horizontal' causation.. That is to say, they conform to materialism's sense of progress 'light-to-heavy', 'simple-to-complex' and so forth. How, then, did particles and elements spill from a cosmic egg? Did chance really generate the cosmic forms or not?

> *'Egg' is an apt metaphor for outworking from within; but development from an egg, far from being a random process, is preconditioned and involves precise codification. And codification, demanding forethought, is a product of mind.*

Bottom-up, grant that photons, quarks, protons, electrons and the other fundamentals rose automatically. They rose from sets of rules spontaneously in place. It *must* have happened in a mindless way.

In physic's realm the automated factory works witlessly but, as we've seen, instinct is a part of archetypal memory. By extension, the behaviours of cosmic body are defined by physical first cause, that is, a metaphysical precursor. Great network or machine are well-known metaphors for cosmic operation. Also root with branching world-tree. Maybe the ancient and organic metaphors of seed and cosmic egg with implication of programmed development have merit too.

So, is there a disconnect? Materialism disallows vertical causation in 'nature' but in all a scientist's work (experimental, observational, technological) allows that it applies. So is his mind 'natural' or 'unnatural'? Because in natural and vertical causation, there is mind behind.

In Chapter 2 the illustration 'Order of an Act of Creation' showed how vertical causation, that is, hierarchical creation works. **Top-down, from Natural Dialectic's spectral view, exterior appearance issues from interior.** Bulk matter follows from its inner, quantum nature. Whence does that nature take its form? Framing ideas accurately yet flexibly is syntax. Syntax, linguistic or mathematical, is convention or a legal framework in which symbols are ordered; its law naturally determines those structures allowed and those not. Thus the *upper syntactic level* (informant phase) acts as a filter through which order is communicated to and from the *lower (environmental, statistical or quantitative) level* (informed phase) of data items. In orderly emergence sub-atomic elements, forces and atoms of physics and chemistry can be construed as basic elements of code. Such code would dictate, through the agents that a study of phenomena elucidates, the way things naturally turn out. Quantum particles have been analogised to alphabetic notes/ marks/ letters of a cosmic language. *Looked at this way cosmos is a Grand, Dynamic and Encoded Text.*

To complete the cycle we need to turn to creation's edge, the world's periphery.

(Tam) Passive Energy

This is the end. *Passive energy* is known as the classical phase of bulk matter. Such **matter-in-practice** is considered to be in locked, exhausted phase. Its 'bonds' include molecular formations, plasmas, gases, liquids, solids and, of course, all study related thereto. It represents projection's furthest radius from Source. The subjects of physics and chemistry deal exhaustively with this level of creation.

For this section let the final word rest with Max Planck. Planck not only first read, recognised and published (in the *German Annals of Physics*) the start of Einstein's revolution, the latter's Special Theory of Relativity. He also pioneered quantum theory, the second pillar of modern physics whose study is sub-microscopic, sub-atomic phenomena - the *matter-in-principle* of Natural Dialectic. Actually, his foresight may have ushered in the next revolution in human understanding which has already begun to focus on the primacy of the 'unscientific' informative co-principal as opposed to its energetic coordinate. He asserted that the discovery of truth can only be secured by a determined step into the realm of metaphysics and, at a lecture in Florence (1944) called '*Das Wesen der Materie*' (The Nature of Matter),[52] said:

"As a man who has devoted his whole life to the most clear-headed science, to the study of matter, I can tell you as the result of my research about the atoms, this much: *there is no matter as such.* All matter originates and exists only by virtue of a force which brings the particles of an atom to vibration and holds this most minute solar system of the atom together...*We must assume behind this force the existence of a conscious and intelligent Mind.* **This Mind is the matrix of all matter.**"

Indeed (The Observer 25-1-31 p. 17), "I regard consciousness as fundamental. I regard matter as a derivative of consciousness."

Perhaps Max Planck would have appreciated Natural Dialectic.

Questions 4

1. What three items compose the hierarchical triplex of physics?
2. Arrange them as the triplex member of a 'stack'.
3. Can a scientist also be a contemplative mystic?
4. What is a holy grail? What is the holy grail of physics?
5. Is physical nature *always* reflex and aimless in its behaviours or is its non-consciousness capable of any purposive complexity or design?

6. In holism, as opposed to physics, what is physical potential?

7. In holism what does gravity mean?

8. Is mathematics a form of physic or metaphysic?

9. What, out of pre-physical void, might 'horizontally' (by knock-on effect) cause a physical universe?

10. What causes some mathematicians and physicists of note to believe that the probability of the physical zone of cosmos appearing by chance is highly remote?

11. If this indicates, as for any complex and dynamic system, a plan then what is a common scientific/ materialistic response?

12. Why is 'cosmic egg' an apt metaphor for cosmogony (the creation of cosmos and its subsequent cosmological development)?

13. Why can we call quantum characters 'matter-in-principle'?

14. Do you think that physical transformations happen, like balls flying around a billiard table, by chance or rigid, mathematically describable determination?

15. With what characters is the text of the world inscribed?

Chapter 5: Biology

Before the 1950s biology was a matter of 'outward' observation, copying, classification and, for about 30 years, some 'elementary' biochemistry (of vitamins, proteins and a suspicion, no more, that a substance called nucleic acid was a central component of all living things).

After 1953, however, things really took off. The structure, operation and 4-letter alphabet of a superb computer language was discovered. *DNA* is to biological form what computer chip is to an AI robot - except far superior in operation in that, like an entirely automated factory, it can cause the reproduction, growth, maintenance and repair of itself and the form it inhabits.[53] *DNA* codes, through an intermediate called *RNA*, for protein. Above are ranged discoverers Francis Crick, James Watson, Rosalind Franklin, Maurice Wilkins and 'runner-up', Linus Pauling.

A founding father of molecular biology was Malcolm Dixon. He was followed by Max Perutz and Fred Sanger who, again here in Cambridge, led the field in sequencing amino acids that compose a protein. In the same lab at about the same time Sydney Brenner realised that *DNA* was a language. Actually it is a digital quaternary (as opposed to binary) code; and, on top of that a double code. It may even be that, as well as epigenetic considerations, some sections of 'inside information' may read productively either way; such palindromic sense would require a very high level of ingenuity to formulate. However this may be, now *DNA* itself can be rapidly sequenced. Soon we shall have a library that includes the texts of all living organisms.

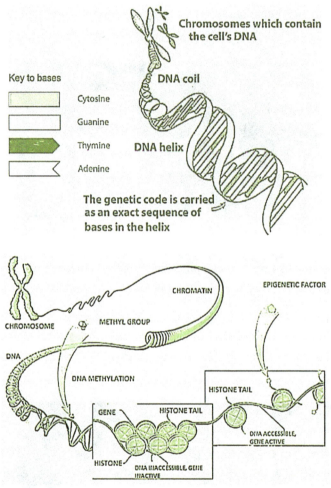

Of course, epigenetics - codes governing codes - amounts to extra informative dimensions.[54] Hierarchically arranged, its mechanisms build a towering, high-grade data structure. *A responsive genome incorporates multiple, overlapping and interactive code in a mode the IT profession calls data compression.* **How can first-class data compression arise without a prior informant?** Ask any software engineer. *Why should DNA and its controllers, packing information just as tight as any hard drive, be different?* **In fact, its chemicals express each bio-program. _DNA_ is nature's chip.** So the basic bio-issue is, unrelentingly, the source of code.

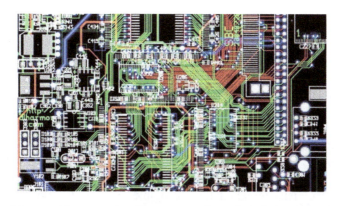

The chip specifies protein. Protein synthesis involves transcription and translation into protein using molecular industrial units, some composed of many subunits. Transcription is a multi-step process employing polymerase and other molecular machines. Translation, at the other end, is also multi-stage; many machines, including multiplex ribosomes, cooperate. **No theory has been proposed for the evolution of the critical, immediate and highly serendipitous linkage of protein with nucleic acid through the medium of such a suite of requisite, complexly-formed and yet precise t-*RNA* translators and corresponding synthetase tools.** Without this series intact nothing could start. So it goes on. *At least 100 different proteins, each coded for and synthesised by the very machine of which they are components, are used to convert DNA instruction into protein product-hardware.* Of the whole system the late Professor Malcolm Dixon, a founding father of enzymology (the study of enzymes) at Cambridge University, wrote, ***"The ribosomal system under which we include DNA and the necessary co-factors, provides a mechanism ... for its own reproduction but not for its initial formation."*** This, since complex, coded mechanisms can't self-code or build. Now, as well as source of code we have to ask how, from the first cell, its programs incomparably output the goods!

On these matters naturalism is tenaciously defended. **The Primary Corollary of Materialism (see Chapter 1)** states that life forms are the product of the chemical evolution of a first cell (see Glossary); and following that, the neo-Darwinian theory of evolution. Evolution is, in this sense of the word, change in the heritable characteristics of biological populations over successive generations. Changes are caused by the main mechanisms of sexual recombination (after sex evolved) and mutation acted on by a filter called natural selection. The creation of life on earth is thus an absolutely mindless, purposeless process.

Biologist Theodosius Dobhzhansky is famous for coining a popular mantra: 'nothing makes sense in biology except in the light of evolution'.

But the actual, iconoclastic fact is: 'nothing makes sense in biology except in the light of information.' You may drag evolution in on information's tail but it is simply a fashionable word used in biology to mean several different things. *In reality, codes and signals run the bio-show.*

In other words, *not* evolution but information is the basis of biology - code that yields irreducibly complex structures and cooperations.[55] So Darwin, seeking an origin of species and ignorant of the biological *sine qua non*, *DNA* codification, asked the wrong question. *It is the natural origin of information needs be rightly explained.* Every cell in every body depends on the successful integration, by program, of complex factors without which it would not work. This is called its irreducible complexity and is tightly linked to minimal functionality (meaning a mechanism must fulfil its promise to the extent that an efficient, acceptable level of performance is achieved). Physical elements and natural forces could not make a cup of tea in a billion years, let alone its drinker.

The issue is not one of religion or opinion but science and logic. Information is, recall from Chapter 2, demonstrably an immaterial factor. It is metaphysical. **Therefore, the basis of biology is metaphysical.** And such holistic perspective (including both material *and* immaterial elements for consideration) renders the notion of chemical evolution logically illogical. *The promissory faith in obtaining a 'first cell' by chance (as opposed to 'chance' guided by great scientific ingenuity both in mind and in the lab) is irrational.* **In other words, the notion of chemical evolution is, in principle, fundamentally and demonstrably absurd.**[56]

***E cellula omnis cellula*.** Only from a parent cell does daughter come. **No exception has ever been found to this rule so that it is called The Law of Biogenesis.**[57]

Moreover, all cells are codified; and codes *anticipate*. Such biological precondition accurately informs the manufacture, maintenance and reproduction of specific, complex cells and bodies. Can a book robotically assemble its own plot? Ask any IT specialist or inventor if non-conscious, aimless nature could better produce his programs or his mechanisms. Time is not the issue. Mind is. To try to obtain a program by chance is, in fact, a most irrational venture. **For this reason, since information is the basis of biology, the *GTE* (general theory of evolution)[58] is, in principle, also fundamentally and demonstrably absurd.**

Why? Don't variations and speciation obviously and continually occur? So too, genetic accidents (mutations) and adaptive potential (Glossary) that may help buffer an organism against changing environments? But such relatively superficial variation (the *STE* or special theory of evolution) is not the same as the initiation of whole systems. And, although still served in a standard evolutionary wrapper, it seems molecular biology is moving towards an emphasis on information (Chapter 2). The objective of The Encode Project,[59] for example, is to identify all functional components of a human genome (genes, chromatin, the *RNA* transcriptome, splicing and, especially, regulatory elements) by their *DNA* coding. The logic is, therefore, that a sequence of bases *means* something; and, as in the case of Champollion's decipherment of Egyptian hieroglyphs, its objective is to translate what the 'bits and bytes' of its program is saying/ doing/ meaning. Meaning resides in purpose; and when you seek purpose you seek mind. To date Encode has discovered, in *DNA* that is already identified as over 80% functional, nearly a million instruments of regulation. Such nexus of control is inevitably precisely integrated; thus, the informative density of its nano-sized but extremely complex 'chip' is stunning. If natural forces could not make a cup of sweet tea in a ten billion years, why should an ARM or INTEL engineer believe that such a 'chip' as *DNA*'s could 'happen'? For scientific materialism it *must* have. But when will mainstream science catch the (metaphysical) information train?

Since holistic logic (Chapter 1) has inexorably brought us to ths point then, in order to sensibly discuss an inclusive alternative, let's suggest a brief but broadened theory of biology. **After that, since variation definitely occurs, we'll check more fully why the popular *GTE* may, although half-true because of its microevolutionary component *(STE)*, be in the whole false.**

Biology in Brief: Information, Function, Structure

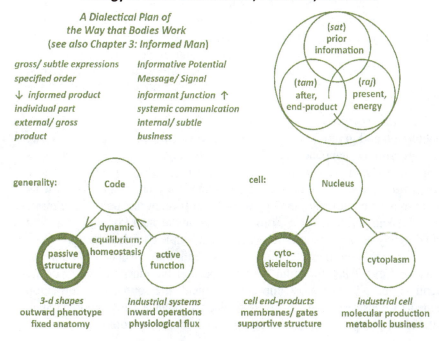

A Dialectical Plan of the Way that Bodies Work (see also Chapter 3: Informed Man)

gross/ subtle expressions
specified order
↓ *informed product*
individual part
external/ gross
product

Informative Potential
Message/ Signal
informant function ↑
systemic communication
internal/ subtle
business

(sat) prior information

(tam) after, end-product

(raj) present, energy

generality: Code

cell: Nucleus

dynamic equilibrium; homeostasis

passive structure

active function

cyto-skeleiton

cytoplasm

3-d shapes
outward phenotype
fixed anatomy

industrial systems
inward operations
physiological flux

cell end-products
membranes/ gates
supportive structure

industrial cell
molecular production
metabolic business

In the green diagram we follow the triplex, dialectical order of expression. *Information precedes.*[60] It is prior and anticipates. Information is the potential, the absolute necessity, for bio-action that consequently issues orderly. Indeed, it was suggested in Chapter 3 that every cell involves sub-conscious mind in the form of its typical mnemone, that is, its archetype.

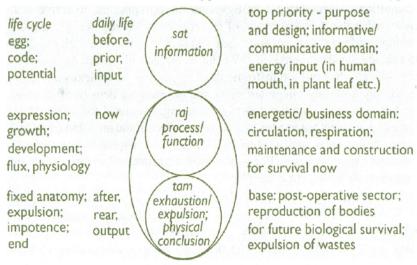

life cycle egg; code; potential	daily life before, prior, input	*sat* *information*	top priority - purpose and design; informative/ communicative domain; energy input (in human mouth, in plant leaf etc.)
expression; growth; development; flux, physiology	now	*raj* *process/* *function*	energetic/ business domain: circulation, respiration; maintenance and construction for survival now
fixed anatomy; expulsion; impotence; end	after, rear, output	*tam* *exhaustion/* *expulsion;* *physical* *conclusion*	base: post-operative sector; reproduction of bodies for future biological survival; expulsion of wastes

Information comes, as previously explained in Chapter 2, in two forms - active/ conscious and passive/ unconscious. It involves both psychological and biological components. **The unconscious, informed case** shows as archetype and soft, bio-logical machinery. This couple includes archetypal instinct and morphogene; reflex balance, called homeostasis, by way of nervous, hormonal and other systems; cybernetic metabolism controlled by preordination in the form of genetic code carried chemically by *DNA*; and muscular organs of action and response. We discussed archetype in the sections on psychology and physics. Regarding physics, the assertion that particles and forces constitute an alphabet, punctuation and grammar of a fine-tuned text is sometimes met with a shrug - things simply happened as they are. *However, considering biology's complex, codified operations you have to shrug this shrug off; the view changes dramatically*. If cosmos is fine-tuned why not, with very intricate and purposive complexity, biology as well? Suffice, at this point, to reiterate that if I asked you for *physical* proof of mind in the chair you are using, you would rightly dismiss the question. Yet even a simple chair has the mind of its designer definitely there. Similarly, mankind may eventually come to understand every last quantum of soft, codified bio-machines but will the chemistry be all? The most important element, the source of information needed to construct them in the first place will have been ignored. Such material-only explanation is, as Polyani noted, in no way complete.

Biology in Brief: Information Plugged into Structured Energy

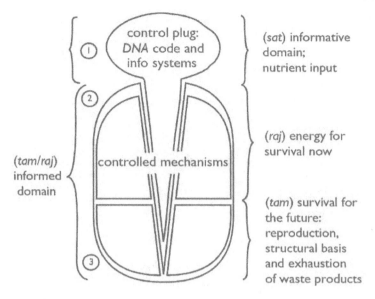

Bio-potential, information, is expressed by specific action. **Energetic actions** provide for survival now. Such energies are delicately informed. The process is one of dynamic equilibrium, equilibration, cyberbernetic balance in accord with pre-set norms. It involves, metabolically, photosynthesis and respiration which in turn promote cell biochemistry, trans-membrane voltages, physiological processes and, on the large-scale, nervous sensation and muscular motion. The character of all function is energetic. All bio-function is coordinated.

Metabolism, being totally information-dependent, works with reference to precise, incoming messages and equally precise genetic response. Such fixed response is indexed, switched and flexibly monitored by non-protein-coding and epigenetic factors, secondary messengers, sugar codes, bio-electrical fields and so on; also by interconnected nervous and hormonal systems. On the conscious, psychological hand, a flux of desires creates a moving set of targets whose equilibration (or neutralisation) is reached in satisfaction. Life, seen this way, is an incarnate flux of order due to information. **In short, organisms are information incarnate**.

Structure, whose character is solidity, represents the outermost, fixed (or flexibly fixed) realisation of shape called phenotype. This 'base domain', made of hard and soft tissue, is energy's container for the expression of (*raj*) internal, orderly flux (called biochemistry). In other words, the peripheral aspect of phenotype both reflects and fixes the shape of inward information and energy. The end-product of structural development, maturity, is reproductive. The cycle starts again.

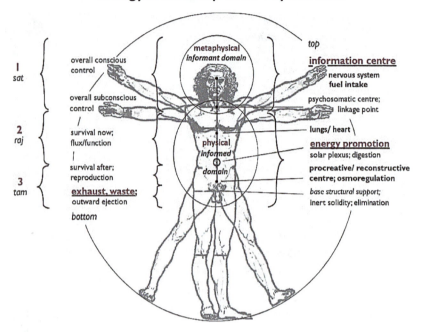

In this view, for Natural Dialectic a cell or multicellular body is primarily information, only secondarily the biochemical forms of molecular biology. And, as we saw in Chapters 3 and 4, the structure of your own body is an expression reflecting the three cosmic fundamentals (see Chapter 1 and Glossary).

We have nuclear super-computing but, in the energy department, the central executive is homeostasis.[61] It is based upon the geometry of periodic action, a circle and its extension, vibration. Vibration round a norm is at the root of biological equilibration, its dynamic equilibrium, its energetic stability or superbly calibrated balancing act. The controls are set for balance; these controls are always, in principle, triplex. They involve mechanisms of sensor, processor/ controller and effector. *All homeostasis, in any cell, must involve these three deliberate components; and each must be codified before it can be built and work.*

Why, in the face of physical entropy, should equilibration be the goal? Why, in spite of nature's fall towards exhaustion, breakdown, garbled chance events and death, should a tight-rope for survival be so evidentially codified? Indeed, why should a group of atoms ever 'want' to survive? Is not the whole cybernetic business of homeostasis for stability of operation? **Cybernetics (the science of automatic control systems in machines and organisms) intrinsically involves anticipation, purpose and a goal.**

Biological energy is, through documented process and specific agents, strictly guided by a process of cyclical control called homeostasis.

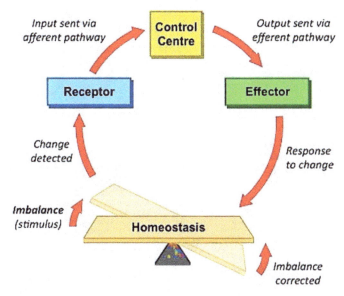

Code anticipates. Signals communicate. Information operates with feedbacks which ensure dynamic systems 'stay in line'. Therefore, to find the causes of 'downstream effects' you have to travel to their 'upstream' cause. *Such cause is not natural selection. Nor is it mutant lesion of a gene or chromosome.*

Hierarchical information-structures are conceptual. In other words, visible characteristics called phenotype depend on molecular action that, in turn, depends on genotypic and other pre-programmed information. *Life is so obviously purposeful in character that genetic chemicals are ascribed all kinds of animistic properties.* They 'compete', 'organise', 'express', 'program', 'adapt', 'select', 'create form', 'engage in evolutionary arms races' and even, lyrically, 'aspire to immortality'.

All biological process is completely informed. And the origin of hierarchical order, cyclical flows of information and integrated function is always purpose. The origin of purpose and its attribute, meaning, is, as Chapter 2 describes, always mind. *We might, therefore, reasonably infer that the basis of meaningfully informative, functional and structural biological hierarchies is mind.* So biology's first, prior set of hierarchies is identified as *informative*.[62]

After *informative* hierarchy (code to protein and chemicals, thence to functions and structures), its second set of hierarchies is *energetic*.[63] As a brief example of informed complexity of mechanism note four illustrations of parts involved in energy metabolism.

Here the top pictures show in three ways one component of respiration, a crucial but very complex molecular 'dynamo' called *ATP* synthase.[64]

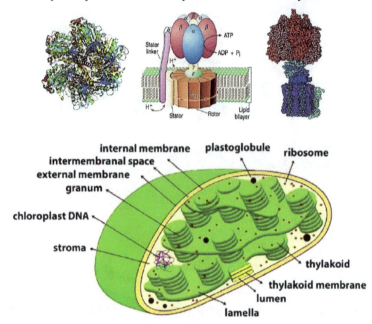

ATP synthase is used to 'burn' the sugars created in an organelle found in plants and phytoplankton called a chloroplast (fourth illustration). Energy metabolism is composed of a duo - photosynthetic build-up and respiratory breakdown of specific molecules. We now know that its complex engineering includes, at the initial stage of photosynthesis, quantum biology as regards the resonant energy transfer of electron excitation in a coherent, most efficient way.

Replication, energy metabolism and reproduction. Whence came the information for all these essential systems? *Bottom-up*, the problem's to convince yourself that, granted a vastly complex starting-point, the computation of cell chemistry haphazardly 'improved itself'. What systems flow-chart could you mindfully build as evidence to demonstrate this mindless possibility?

Top-down, programmers know that, from a main routine, switches branch to sub-routines and, when a sub-routine is done, the process cycles back to start again. **These routines are modules**. This conceptual character of algorithm, this repetitious use of switches and blocks of modular code is just what coded bio-systems show. It is how nature's life-forms, full of reason, always work.

One would, therefore, predict research will more and more reveal signs of bio-logic to the point that, in 'live' computing, such complex

permutations, integrated combinations and hierarchical sets of regulation will further squeeze then nullify the notion that celled systems ever cropped up accidentally, that is, evolved.

A computer is a mind machine.[65] *On this basis it is established that the cybernetic operations of a cell, an object as thoroughly material as a computer, superbly meet the criteria that pass it as a mind machine.* A cell is a mind machine. Its instruction manual is a program written up as 'genome'; and its systems hold semantic meaning as modules or entities of bio-form like, for example, you.

Various tailored sub-routines are called from a Main Routine. Suites of such modules combine as permutations around which different bodies are, under the coordination of such an Archetypal Master Routine, expressed.

Molecular biology's great strides are also heading straight towards the notion of an archetype. A similarity of genetic instructions is found to extend across a great variety of forms. For example, a high-level developmental complex of modules (called 'homeotic genes', *see* Glossary) may code for outline body-plan in very different organisms such as human, duck or fly. Or it may call subroutines for the construction of completely different kinds of eye - a normal gene from a mouse can replace its mutant counterpart in the fly; now an eye, a fly's eye not a mouse's, is produced. In other words, such genes act as 'go-to' switches in an archetypal program of development.

One could add much more but we now turn to a brief critique of the biological theory of evolution in terms of three tenets and five ingredients: adaptation, speciation, natural selection, mutation, sex and development.

Perspectives on Three Central Tenets of neo-Darwinism - a Tabulation		Bottom-up	Top-down
✓ true ✗ false			
①	Abiogenesis/ chemical evolution	✓	✗
②	Variation (so-called microevolution) by mutation, adaptive potential and natural selection	✓	✓
③	Transformism (macroevolution) by mutation, natural selection or any other suggested means	✓	✗

A half-truth is the most difficult to unravel. Elements of Darwin's theory are, of course, agreed by everyone. Everyone agrees that variation-on-theme continually occurs. Such variation is the result of sexual reproduction (which contains its own in-built lottery called meiosis) and, deleteriously, by genetic mutation. ***Top-down, it can also be the product of adaptive potential.***[66]

Remember what was said at the start of this Chapter? **The basis of biology is information. *Therefore, Darwin asked the wrong question. With what we now know the question does not concern an origin of species but, fundamentally, the origin of information.*** Nevertheless, let's take a look at speciation.[67]

Vehicle	Body (vehicle of life)
1. carriage inside or outside vehicle	domain prokaryote/ eukaryote
2. type of propulsion	kingdom e.g plant, animal
3. motion in air, on land or in water	phylum subphylum e.g vertebrate
4. land: motor car, lorry, van etc.	class: mammal
5. car	order: primate
6. make: Porsche	family: *Hominidae*
7. Carrera	genus: *Homo*
8. species of Carrera e.g Cabriolet	species: *Homo sapiens*

Carl Linnaeus, the founder of modern taxonomy, thought a 'species' was a group whose members could interbreed. Darwin, as opposed to Linnaeus, preferred to 'plasticise' his definition of species to what he saw as its cause - gradual change. He wrote *"I look at the term species as one arbitrarily given, for the sake of convenience, to a set of individuals closely resembling each other, and that it does not essentially differ from the term variety."*

However, Darwin not only regarded varieties as incipient species but also proceeded to *extrapolate* on the hypothetical principle of unlimited plasticity. He *did* consider this, a process called micro-evolution with which everyone agrees, to be a minor stage of macro-evolution. In other words, he **guessed** that variation was a progressive rather than a constrained, cyclical process. The issue is, therefore, one of plasticity, that is, the actual extent of variation.[68]

Such extrapolation, variation-*without*-theme, formed the very basis of his theory. Through materialism's prism life is, naturally, progressive from a simple start. This, unlimited plasticity, is unquestionably the current scientific mind-set. Yet, we'll see, even the notion of a 'simple' start (called abiogenesis or chemical evolution) is fraught with intractable problems.[69]

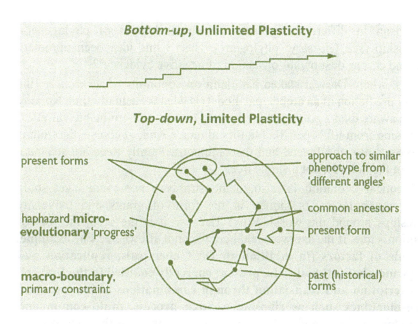

Bottom-up, Unlimited Plasticity

Top-down, Limited Plasticity

present forms

approach to similar phenotype from 'different angles'

common ancestors

haphazard **micro-evolutionary** 'progress'

present form

macro-boundary, primary constraint

past (historical) forms

In practice, when genetics decisively demonstrates the limits past which a mutant does not survive it also demonstrates conclusively, through myriad experiments (e.g. with *HIV, E. coli* bacteria, fruit flies, malarial parasites, flowers and mice), that the macro-evolutionary principle of unlimited plasticity (or unbridled extrapolation) is incorrect. Ask a dog-breeder if he has bred other than a dog. Ask him whether anyone has ever, by intelligent selection, pushed canine elasticity into the production of other than crippled or still-born, monstrously deformed dogs. No-one has made another type of organism from a wolf. This demonstrates a limited plasticity.

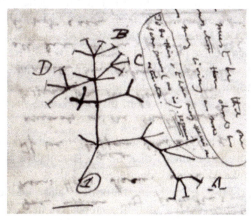

Darwin's inspirational doodle isn't right. The central assumption of his tree of life is that homologous molecular or morphological patterns will, by comparison in different organisms, illustrate degrees of phylogenetic relationship (*see* Glossary: phylogeny). Hasn't this idea been uprooted, sawn and cast as dust by an 'onslaught of negative evidence'?[70]

Thus, where Darwin staked his claim on continuity, *discontinuity* (the primary prediction of an archetypal model) is what we actually find. Nobody is more aware than a *palaeontologist* that Darwin's 'interminable varieties' are missing from life's petrified family album.[71] And, of course, the truth is that types[72] are ring-fenced and discontinuous. Family trees are separate. Life's in fact a forest not a single tree.

Chemical evolution is, also, demonstrably a non-event[73] yet study attracts much taxpayers money in the form of grants and university tenureships around the world. Why? Because it is the seed, the root of evolution's tree. If the axe were laid here why not elsewhere? **For example, a couple of factors (in addition to *DNA* chemicals, replication and energy metabolism) had to be present correctly codified in the first cell - transcription and translation through combinations of enzymes and micro-machines such as ribosomes. Such process, multi-component, multi-stage and self-coded-for, must be working from the very start of bio-informative operations.** In this case, why is the rational notion of an archetype so unintelligent? It is only materialism and, especially, its subset of scientific atheism that *need* to damn it so.

The human genome could be stored on perhaps 1.5 gigabytes of space. Its 250,000 or so pages of close-written lines of letters would compose a book weighing 450 kilos. This code needs be accurate. At the other end of biological scale biologist Craig Venter defined a minimal genome by removing genes, one by one, from the already very small genome of a *Mycoplasma* bacterium. If the organism survived he discarded the gene until he was left with a minimal genome of 473 vital genes. But symbiotic *Carsonella rudii* has certainly lost the bare necessities of life since its present genome of 182 genes (~160000 'letters') is insufficient to replicate or transcribe and synthesize protein. A viable analogy for free-living *LUCA* (hypothesized but uknown 'Last Universal Common Ancestor') might be ten times that amount, that is, 1.6 million letters (the size of a 600-page book) in the precise order of a program that will create specific molecules and shapes.

Next, are natural selection and mutation up to the task of sufficiently 'improving' the survival chances of a highly speculative first cell? First, historically, let's take Darwin's genetic editor called natural selection.[74] Edward Blyth (1810-73), a keen naturalist and 'Father of Indian Ornithology', published essays that first appeared in The Magazine of Natural History in 1835, 1836 and 1837. *These, which were read by Charles Darwin who afterwards corresponded with him, introduced the ideas of a struggle for existence, variation, natural selection and sexual selection.* These four are central Darwinian tenets and we might ask why, since in large part Darwin's work was based on Blyth's ideas, the former hardly acknowledged the latter's 'intellectual copyright'. And Alfred Wallace (1823-1913), co-founder of the theory of evolution and whose paper 'On the Tendency of Varieties to Depart Indefinitely from the Original Type' preceded (1858) and precipitated publication of Darwin's 'Origin of Species' (1859), noted that natural selection not so much selects special variations as exterminates the most unfavourable. Such extermination is, in fact, a stabiliser; 'fit' traits blossom, 'unfit' wither as an organism's own ecology (its niche) dictates.

So let's treat editorial selection for what it actually is and, at the start of our inspection of its 'mechanism', post three *caveats*.

The first, Blyth's, is that natural selection is solely a process of elimination. It involves the disappearance of 'unfit' organisms. **In short, deletion is not creation. A corrector, not a selector, Natural selection originates absolutely nothing at all.** It is, simply, a fateful finger hovering above a genetic delete button, no more than a name given to the lucky survival or unfortunate death of an organism or group of organisms in a particular environment.

Secondly, the effect of natural selection is, in the case of speciation, to have *reduced* the genetic potential of an original gene pool. Alleles are deleted or a pool split into separate populations. Information is not gained but lost. *For evolution, which needs not information* ($\downarrow$) *loss but* ($\uparrow$) *gain, this is entirely the wrong direction.*

Thirdly, it is false to conceive of natural selection as a kind of 'ratchet' that holds on to 'an extrapolation of improvements' at any level from nucleotide through protein to whole body shape. This is because without a prior plan to work towards there is no way for intelligence, let alone total lack of intelligence, to 'discriminate' a 'good' apart from 'bad' move as regards some novelty. *Indeed, it will breed out unwanted, nascent or non-functional characteristics.* **Only once an organ or a functional system (such as a beak or eye and associated factors) is complete and fitly working can natural selection act on trivial, accidental variation to that system. In a phrase, the 'mechanism' explains survival not arrival of the fittest.** It weeds the weaker but cannot create the fitter; it's as creative as a kitchen sieve.

In short, this mindless editor confers not transformative change but stability by maintaining wild-type pedigree. Far from being a grand 'law of nature' natural selection is a trivial observation - an organism born defective does not reproduce. Its truism states the obvious. 'Weaker die, stronger live'. 'Who survives, survives'.

To say that 'nature' is the mother of invention is to put it mildly. What, therefore, about the aimless mechanism that provides 'designs' that natural selection 'excellently' hones?

Mutation.[75] In 1865 Gregor Mendel presented his Theory of Discrete Units of Heredity. This was picked up by the Royal and Linnaean Societies in London and (since 'adamantine particles' are inflexible and anti-evolutionary) ignored until, in 1901 de Vries discovered a way out - **mutation**. Henceforth random mutation became the innovator, the creator on which natural selection might, by killing, work. **Materialistic faith is vested in a core of unpredictability, a central lack of reason - mindless chance.** A scientific G.O.D, no less! A Generator of Diversity! By philosophical necessity current science takes its chance on chance. Order came about by chance. No telling how exactly, just vague imprecation. The story's scientific *aide-de-camp* is probability. Nothing is, perchance, impossible; the sole impossibility is that such a story is impossible. Materialism's Lady Luck, however weak, is elevated to almighty creativity, creator of life forms.

Of course, chance does impact life. In a few cases and contexts genetic mutation may help a species survive and thrive. Lucky chaps! But if mutation simply breaks or loses information from a gene then it *devolves* an old form not evolves a new. **At best it generates predominantly damaged variation on a predetermined theme**. Today's mutations are, arguably without exception, construed as causes of disease or abnormality while, in the past, they are proclaimed creators of the tree of life!

Nevertheless, mightn't mutations, if you found millions of 'beneficial' ones, transform one type of body to another in the way accretive evolution needs? *'Beneficial mutations' (called BMs) are to genetics as 'missing links' to evolutionary palaeontology - critical.* **Not just The Primary Corollary but The Primary Axiom of Materialism (Chapter 1) and its whole panoply of philosophical, political and sociological speculation, not to mention the paradigm of modern science, hangs on their slender thread.** *The whole of secular academy depends, for its verbose existence, on this evanescent gleam of hope, a key to unlock all of life's diversity that's a 'positive' or 'beneficial mutation'. Can BMs really rise to such occasion? Everything depends on this.*

In fact, any *BM* would be invisible at the level of a whole organism and its small 'advance' overwhelmed by neutral or deleterious mutations long before exact serial and parallel chains of cooperative *BM*s needed for any novel biological system could 'evolve'.

Not knowing where you're going is a fundamental problem too. It applies to 'junk', 'neutral' or any other kind of *DNA*. What should a first or following *BM* be? One can only be defined within a preconception. What is 'good' or what is 'bad' is only so when valued in anticipation of that end. How can you even define a *BM* if you lack direction? The atoms of a system do not understand it. Mindless evolution's natural selection doesn't know. And bio-analysts rarely, if ever, detail the stepwise algorithms such complex but mindless construction might demand.

But if *serial BM*s pose an 'innovation problem' the diagnostic worsens. In reality, if an engineer wants to radically revise, adapt or innovate part or all of a system then myriad precise, interlocking adjustments will be needed before it works. **Similarly, evolution would require not serial but multiple *parallel*, synchronous, cooperative *BM*s.** In fact, if there's no such thing as target how, except by wishful thinking, can 'progress' gradually mint a system with its integrated working parts? **With such lack of transformational logic bang goes materialism's irrational *PCM* and therefore bang goes *PAM* as well!**

The fact is that *DNA* represents, effectively, an organism's program. It has, however, never been shown that a coding system and semantic information could originate by itself through matter. IT's information theorems predict this will never be possible. Yet the basis of biological program is code. Code is always the result of a mental process. If code is found in any system, you might conclude that the system originated from concept, not from chance - especially if that code is optimised according to such criteria as ease and accuracy of transmission, maximum storage density and efficiency of carriage (such as electrical, chemical, magnetic, olfactory, on paper, on tape, broadcast, *DNA* etc.); and if, above all, it works and orderly instructions are unerringly responded to. Code and incoherent chance are chalk and cheese.

Randomness of any kind is reason-in-reverse. Whatever is encoded is intentional. *A coder takes no chance. Randomness is eliminated.* By definition, mistake or randomness degrades information; and the job of any editor (programmed *DNA* editors included) is to eliminate interference, 'noise' or mistake. **Chance neither creates nor transmits information. On the contrary, accidents always (unless accidentally reversing a previous degradation) degrade meaning and, by degree, render information unintelligible**.

Does *DNA* really operate like a digital computer program?[76] There are close resemblances. And every systems engineer knows that a change to his construction (whether accidental or deliberate) either brings things to a grinding halt or causes ripple effects that, unless properly balanced and calibrated with cooperative parts elsewhere, degrade the functional intention of his design. Similarly with software. Random changes without reason or systematic realignments have, at best, no effect or, at worst, bring catastrophic failure. They never improve it! This is exactly what we find in the case of biological mutations. They are not a good choice of creator to underpin the theory of evolution and its materialistic creeds. Indeed, they intrinsically pronounce it incorrect.

Yet, from ignorance, randomness replaced, by design, design! It was touted throughout the 20th century that we knew such bio-programs whose operation we still hardly understand evolved by chance mutations acted on by natural selection. Chance mutations *'must have'* caused the code with its adaptive, switching systems but, after this appeal, it is never explained exactly how (in what combinations or algorithmic steps) or why such mutation was, although at that stage neutral or harmful (i.e. useless or worse), on the way towards some 'improvement'. **This is not science. It amounts to the multiple invention and repetition of just-so stories; and is, in this sense, an argument not from code but guess and ignorance.**

Yes, variation-in-action certainly exists. All agree upon the limited plasticity of micro-evolution. And on adaptation to fresh circumstance. Could not, however, an anticipatory buffer called **adaptive potential** (Glossary) be written into genome and its epigenome so that it would not be mutation but a flexibility of program that produced coloration changes, different beak sizes and so on? Various mechanisms have been suggested, none of which , however, come near to explaining the *innovation* of any integrated, functional programs, organs, systems or, beyond variation-on-theme, transformist macro-evolution.

In science you experiment. Can you test evolution in the past? In his book 'The Edge of Evolution' Michael Behe demonstrates that nature has empirically tested Darwin. Do you want numbers showing how, by gradual mutations, life might step-by-step evolve its family tree? *HIV, E. coli* and malarial parasites satisfy the numbers game. Virus, bacterium and eukaryotic cell have reproduced, mutated and should have evolved through sufficient generations with sufficiently large populations to indicate whether neo-Darwinism's engine, random mutation, can bear the weight of a theory that would have it gradually innovate parts and body-plans by gradual but cumulative, useful steps - or, buckling, is crushed by numbers.[77]

In short, we understand that genetic information is at the root of biological form creation and maintenance but not how the program 'knows' how to build objects and events that make precisely the right parts at the right times in the right quantities and places. We do not know exactly how this automated factory program generates development (which involves anticipation) of any tissue, organ, system or whole body. When we do we will be able to say that we thoroughly understand this or that machine in all its aspects. But will this mean, as a materialist might claim, that we completely understand?

As Michael Polyani argued, machines are irreducible to physics and chemistry. *They are irreducible because they involve immaterial purpose, the stepwise development of a plan of implementation, a directed cohesion of working parts and, of course, the thoroughly non-material anticipation of an operational outcome.* For example, to completely analyse a bicycle does not mean you have completely understood that form. You need to include its purpose, conception and technological development, that is, the vertical causation of mind in it. Simple physico-chemical analysis would never by itself obtain more than a fraction of the whole truth.

Seen in this light, micro-evolution is a prejudicial word for variation-on-theme, biased because it implies the existence of an extended process for which no hard evidence exists. **The reverse. The extension called macro-evolution is a theoretical phantom. In this view, evolution-in-action is simply variation-on-predesignated-theme**; this variation is always constrained by working systems already in place; and it is either coherent, due to in-built genetic potential or incoherent by neo-Darwinian mutation. ***Variation proliferates; the special theory (STE) is right. But the general theory of grand macro-evolution (GTE) is not.***

Energetic matter knows no future. It cannot program, plan or anticipate. Waste is exhausted. Yet, informant egg to informed adult, *reproduction* shows anticipation. Target. Purpose. There is exhaustion of waste materials but also, by reproduction involving the lower quarters, expulsion of fresh bio-forms. Reproduction[78] defies entropy and death. It means the type survives.

There are three main ways that, with many detailed variations, organisms reproduce. To save time we take, of these, sexual reproduction.[79] The biological point of sex is variation-on-theme. It is neither to add nor subtract but to shuffle an organism's cards into new permutations and thereby deal new hands in an old game. In this case two become a different one. Sex involves clear anticipation, targets and complex programming - not features normally attributed to mindless matter or to chance. How did it arise?

The Archetypal Polarity of Sex 1

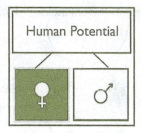

If you wanted, theoretically, to design complementary sexes you might opt to specify each module using thousands of genes; *alternatively, and far*

smarter, you might conceive a main routine which included a 'gender switch'. Each type of life form that exhibits sex is conceived of as a neutral whole divided into polar male and female sexual halves.

The sexual archetype (<u>not</u> a common ancestor) and its *DNA* coding are fundamentally hermaphrodite and incorporate both male and female potential. Such hermaphroditic concept's switch would trip a male or female line; it would call gender-informed sub-routines. In principle, therefore, at a mere flick the balance would be tipped. A hierarchical cascade of emphasis would, either way, ensure dimorphic forms occur. **This skilful concept, this consummately programmed switch is precisely what we find in bio-practice.** From the same cells grow, in each gender's case, male and female parts. For example, sensibly perceived, male breastless and female breasted nipple are, as penis and clitoris, examples of the differential expression of the human archetype. Indeed, dimorphic *and* hermaphroditic (but never multimorphic) sexual algorithms are found in plants and animals. From hermaphroditic potential derive uni-morphic hermaphrodite, di-morphic male/ female, alternation of generations (alternate sexual and asexual phases in a life cycle) or various forms of sequential hermaphroditism (change from one sex to another or one sex to hermaphroditic form within a life cycle).

The Archetypal Polarity of Sex 2

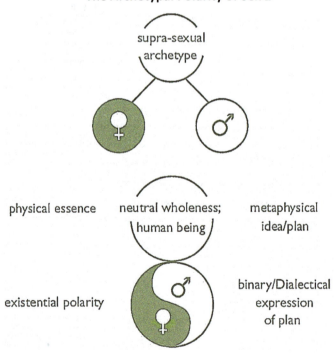

You know for certain, therefore, the same program can by switch deliver, for example, sperm (including the set of over 600 proteins that compose its eukaryotic flagellum) and egg (with her own specialities). Of course, genetic switching *aberrations* (e.g. intersex, abnormal hermaphrodites or trans-sexual tendencies) may occur as well but, normally, sex is implicit in a neutral archetype's potentiality.

And as for eggs, did an egg precede its adult?[80] You can't see a memory but have you ever seen a physical idea? Ideas are potential; from potential outcomes are evolved; an egg, packed with symbolic information, is as near to 'natural idea' as you will get.

Did an adult precede its egg? Did fruit precede its branch? We noted that reproduction looks to the future; and thus sex *anticipates*. It is a conceptual process engaging machinery of irreducible complexity to achieve a target - generation after generation. Survival of the kind.

Egg and Adult Together. What was noted in Chapter 2 bears repetition. **Wherever an apparent 'chicken-and-egg' situation crops up the puzzle is resolved by the introduction of purpose, design, information and mind rolled into one - teleology.**

As far as chicken and egg are concerned, therefore, the simple answer of an information technologist is that neither came first. One did not precede the other. You make them together with the same object in mind - in this case a self-reproducible biological information system or, in other words, a living being. **You are one half of a metaphysical idea called 'human being'.**

Meiosis, which underwrites the sexual reproduction of some single-celled and nearly all multicellular organisms, bears the hallmarks of a choreographed routine designed to extract maximum variation-on-theme at minimum cost in labour and materials.

Meiotically generated gametes then, by means of and within a highly specified and complex context, fuse; and their offspring, a single-celled zygote, unfolds according to complex but codified, specific algorithms. A hierarchical developmental archetype eventually realises its goal, its reason - re-creation of the next adult generation.

At this point we can ask how an engineer might design a self-reproducing machine and ask if nature has preceded him in his intelligent logic. How might the least demand be made on the tissue or strength of a parent while at the same time encapsulating its potential? A brilliant, optimally economical idea would be to reduce the parent to a single cell and then, from the symbolism of this cell, build up a new adult. This is exactly what happens in nature. Each individual adult is 're-potentiated'. Its body is reduced almost entirely to a symbol, a directory, a coded book of what might be. Its spring is re-compressed into the top-level potential of an egg or sperm.

The Order of Development:
Developmental Stereocomputation

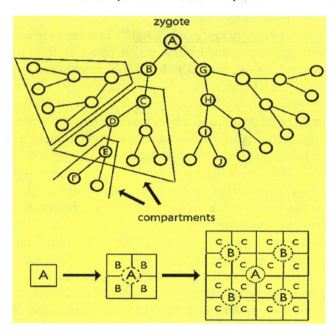

zygote

compartments

Hierarchical Control

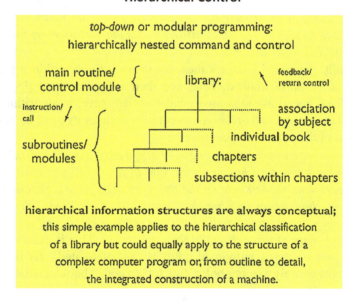

top-down or modular programming:
hierarchically nested command and control

main routine/ control module

library:

feedback/ return control

instruction/ call

association by subject

subroutines/ modules

individual book

chapters

subsections within chapters

hierarchical information structures are always conceptual;
this simple example applies to the hierarchical classification
of a library but could equally apply to the structure of a
complex computer program or, from outline to detail,
the integrated construction of a machine.

Developmental biologists uncover switch-arrays and associated, complex, hierarchical cascades of regulation.[81] They reveal top-level 'master genes' or 'tool-kit factors' that determine which parts of a body will develop out of others. These blocks of levers clustered as a signal box that rules developmental lines remind you of a railway running from a start to terminus. With main and branching sub-routines they remind you also of computer software structures. Regulatory logic-clusters are found, it will predictably transpire, in all multicellular animals. For example, in four distinct blocks in vertebrates (called *Hox* gene clusters) each gene is responsible for triggering a cascade of sub-routines that will supply the right materials in the right place at the right time to construct a given segment of the animal in question. Obviously, therefore, a *Hox* gene for a worm will define (or, in computing terms, call) different or deeply differentiated organs (sub-routines) to ones for a fly, a horse or a human. In other words, the same genes can be responsible for initiating the development, in terms of systems or organs, of different outcomes. In one instance a gene may specify for a tail, a coccyx or the rear part of a fly, frog or grasshopper; and the gene that triggers development of your eye would, if transferred to a fly that lacked it, cause its blindness to be eyed - with a fly eye not a human one, of course!

Yet, furthermore, it has been found in fruit flies that, although heads still remain heads and tails tails, genes specifying basic head-tail orientation in early embryogenesis can, instead of being conserved, differ between species. Such 'top-level' genetic exceptions appear, *prima facie*, to break the rules of Darwinian inheritance. If, moreover, they turn the idea that specific homeotic genes are responsible for typical shape on its head, then what exactly is responsible? Surely not a supra-genic morphogene?[82]

In short, a dialectical bio-hierarchy is not horizontally caused by genetic accidents through time, that is, by a successive 'origin of species'. It is, instead, the product of vertically informed genetic program (with its preconceived metabolic and physiological subsequences) designed to develop and maintain organic structures. If there is objection to this idea it is not scientific but due to the philosophical prejudice of naturalism and its consequential, mindless author - chance.

For example, how does chance evolve a path that targets goal? How, entirely ignorant of end-game, did as-yet-useless intermediates survive? Missing links die of incompetence thus how did any metabolic pathway prophesy its own construction or feedback control; how (as, analogously, in the logical, consequential proof of a mathematical theorem) did thousands of correct steps on the path to reproductive adulthood arise by accident? Evolutionary 'must-have-somehow', 'just-so' stunts are pulled continually except, with metamorphosis, the bluff is called spectacularly. Butterflies have always thrown Darwinism, in an arm-lock, in a flap and on its back.

Back-to-front. The transformation from egg to beetle, bee, fly or, more strikingly, a butterfly illustrates a pattern of development that defies a gradual, practice-to-principle evolutionary explanation.

Entirely different-looking phases, each perfectly formed for function, serially erupt. First *egg*; then, for a 'childish' *caterpillar* to become an adult moth or butterfly, it eats and moults. It keeps moulting exoskeletons (which are flexible like cat-suits and yet give it shape) for larger ones that form folded underneath the smaller outer sheath. At the last and largest size skin is shed by delicate manoeuvres revealing a cocoon, a *chrysalis* in which the future hangs. Then caterpillar body parts dissolve and build again into a butterfly called the *imago* that, emerging after several days, inflates its lovely wings by pumping blood into their veins.

Could humbly waiting as a caterpillar eventually evolve a program (saved as what are called 'imaginal discs') for pupal development? In what nascent cocoon could enzymes 'know' how far they should dissolve a caterpillar's parts before re-building to transfiguration called a healthy butterfly? How long did some ancestral pulp hang round inside a chrysalis (that came from who knows where) until a suite of chance mutations magically (how else?) re-programmed 'mush' into the concept of winged flight? How, in fact, did pupal death rise straightway (not even in a generation) to a form in which reproductive type-butterfly appears? One must ask also how, without the benefit of plan to reach anticipated adult form, serial immature phases took hundreds of millions of years to accrete. *It is noted that before any organism could 'create' any new stage of development it would have first to reach its present stage's adult limit then add that new stage.* This is because it has to reproduce to create the fresh, 'advanced' offspring. A reproducing chrysalis, caterpillar or top-level, coded egg? Such back-to-front order is patently absurd.

From principle to practice, conceptual information to biology - imaginal discs, *Hox* genes and other features simply demonstrate the *top-down* venture that, at climax, claps on stage a butterfly.

Butterflies are symbols of nemesis. At the clap of silent, fragile wings Darwinism logically dies; a giant is slain and at the same time flutter flags of life's innate, original intelligence.

Signs of obvious anticipation always flatten evolution. They squeeze the theory's time to death and thus compress it to impossibility. How can forethought be by chance? When is a plan not a plan? Is concept the same as lack of it? Biological evolution entirely *lacks* concept or target yet its theory equally lacks any serious explanation how, through many specified and integrated stages serial *and* parallel, the evolution of development '*must have*' occurred. Is this 'rationalism's' finest hour? **The fact is that all instances of metamorphosis (of which development in general is one) are another Darwinian black box.**

Sex, metabolism, metamorphosis and morphogenesis are four anticipatory and therefore conceptual processes. **The fact is that the evolution of development is as much a black box as the evolution of any of them. Their bio-logic hammers nails squarely into at least four corners of Darwinism's and thereby materialism's coffin.**

Questions 5

1. What was the 20th century discovery that prompted a revolution in biology?

2. What is the basis of biological form?

3. If the basis of biology is information in the form of code, is this basis physical or metaphysical?

4. Is chance compatible with code?

5. How did a nucleotide first appear? What proof is there that polynucleotides chemically evolved, that is, appeared spontaneously due to local conditions?

6. DNA controls, through a complex process, the shape of proteins. Could a protein factory, needed from the start of life on earth, pop up in water or on land in spite of raw energy (in the form of, say, radiation, heat or violent current), in spite of the lack of it (say, frozen circumstance) or in spite of entropy?

7. Does entropy (tendency to lack of order) appear to help the evolutionary case?

8. What does the Law of Biogenesis' say?

9. How did Darwin construe 'species'?

10. What was Darwin's great guess? Was it right or wrong?

11. What is the real issue here? Is it 'non-progressive (blind) progress' through millions of years or not?

12. What actually runs the bio-show?

13. What do symbolic instructions do that any uninstructed process cannot?

14. Why is the operation of codified instruction hierarchical?

15. Biological forms are hierarchical. Orders are passed down a chain. Structures are developed in preordained steps. Can a hierarchy spontaneously appear?

16. Biological forms develop and survive, for a while, against the vicissitudes of wear and tear. They do this through a process of dynamic equilibrium. What does this mean?

17. What is the biological process of maintenance called?

18. Could such a process, critical to survival in that it inhibits chemical, structural and functional breakdown, build itself?

19. Why is hypothesis of chemical evolution said to be 'irrational' and the theory of evolution 'in principle, fundamentally and demonstrably absurd'?

20. What is the triplex biology suggested by Natural Dialectic based on?

21. Write this triplex in the form of a stack.

22. What are the informative systems in a cell?

23. What are the main supra-cellular informative systems in an animal body?

24. Could an informative system build gradually and piecemeal in a plant body?

25. If bio-information systems are hierarchical and goal-oriented what about bio-energetic systems? How do chemosynthetic and photosynthetic systems work?

26. Charles Darwin did not understand that the basis of biology was information, information in the form of code. As a result he

asked the wrong question about the origin of life. What was this question?

27. Do variation, natural selection and mutation (unknown to Darwin but now the engine of theoretical evolution) occur?

28. What is the difference between limited and unlimited plasticity?

29. Is there evidence that plasticity is unlimited?

30. Doesn't the fossil record prove Darwinism is correct?

31. What about 'missing links'?

32. What is a simple way to define randomness?

33. What part does randomness play in biological form and function?

34. What is a beneficial mutation?

35. What is natural selection?

36. What did Michael Polyani FRS, chemist and philosopher, have to say with respect to machines? Was he right?

37. All machines anticipate their goal. Can you think of two especially obvious examples of biological anticipation?

38. How do you think sex evolved?

39. Calculate how many organs have to co-evolve before sex can achieve its goals.

40. What is the holistic suggestion regarding sexual origins?

41. What is scientific holism?

42. Can you design the reproduction of a multicellular organism?

43. What is stereocomputation?

44. By what steps did the evolution of development occur?

45. How might metamorphosis evolve?

Chapter 6:   Community

The system of Natural Dialectic is, as much as being an abstract reflection of the way things are, an application program. This program generates a template for both involuntary, instinctive and voluntary, chosen behaviours. It organises the pattern not only of 'hard science' and biology but politics, law and religion; it guides the aspirations of education; it is hard-wired into the humanities and, as such, is expressed in the very fabric of individual and social life. Its consequences, especially moral and psychological consequences, involve everyone. How?

We're going to work towards a Unified Theory of Community.[83] **From a bottom-up perspective humans are animals and mind evolved as a strange function of brains. Top-down,** *however, the first compass of community is universal and absolute.* It reflects a structure of creation whereby all things, psychological and physical, descend from an Absolute Source. In this sense alone is everything connected or, as some aver, 'is one'. **Such Absolute Community of Essence and Existence is illustrated in this blue diagram.**[84]

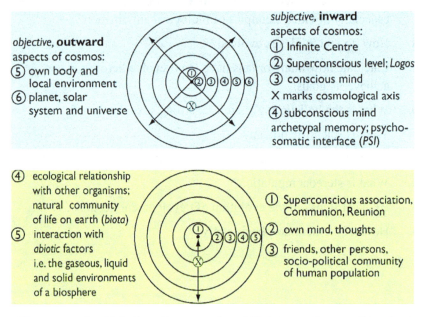

However, the Relative Community of Existence from a Standpoint of X (green diagram) represents our own cosmological bearings.[85] X is the eye-centre, a point on your forehead between the eyes. From here social circles radiate in *both* directions - in circles outward towards a variegated world (3-5) and inwards, through mind, towards our Single, Nuclear Axis

(1). Thus such motion may, most paradoxically, culminate in association with the Source and Centre of the universe.

Thus, *internally* there exist the symbolic yet most real relationships of mind. They involve other persons, bio-forms, events, objects and the paraphernalia of a life spent learning. This is the experience of your mind-world. You.

Externally you descend from X through physical interaction with your environment. The rings spread first through your nervous system (which is the closest physical associate of mind) to the rest of your skin and bones and their neighbourhood. This is your space-time location. It includes other biological bodies to which you physically and, more or less easily according to type, psychologically relate. If you restrict these to the human type then you involve the dynamic of family, friends, neighbours, interest groups, nations or the international scene. You take a social part, which may involve institutionalised politics, law and religion, in your community.

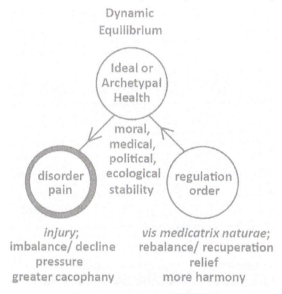

This diagram represents a summary of social dynamics in terms of Natural Dialectic. *At broadest, you can think of life-forms (biotic factors) in terms of individuals, populations or communities of different kinds of organism.* **This positive, inclusive perspective is called <u>ecological</u>. *Thus ecology is about your 'wider body'.*[86]** Such body includes a community of organisms (*biota*) that inhabit the surface region of our planet. This region, called an ecosphere, comprises living and non-living parts. The latter descend, as the rings convey, to include non-living elements (*abiota*) such as sources of energy, climate, ocean, water, mineral cycles and the soils of earth. The compound amounts to a stage on which, in different scenes, various players act. An ecological play is dynamic. In such a network the health and behaviour of each part affects its whole.

Firstly let's very briefly survey non-living factors.

We're here. Therefore at least our planetary zone is habitable. But life needs an endless supply of liquid water and thus strict, natural thermo-regulation. Our thermal generator is a 'dwarf main sequence star'. Our lives are hung upon this lucky star. Precise strength and character of the four basic physical forces keep it burning radiantly. Not only earth's distance from the sun but also its unusually circular orbit slung on a fixed radius and a specific kind of rotation round its own axis is each exactly felicitous.

Pure energy (sunlight with harmful frequencies deflected or deleted) falls on gas. Earth's primary dynamic, sky, consists of a concentric suite of atmospheric shells. Each, like a membrane, offers its particular aegis to the life within. The living planet floats, egg-like in a white of air, within the warm, deep womb of solar influence. You might construe the stratosphere and ionosphere as buffers, membranes, even subtle skin.

Nested lower down the suite of air-light shells you find fluids of fertility - mists, rain, rivers and the oceans. These, lower atmosphere and ground water, assume the blood-like role of convectors, radiators and conductors. The amount of water on the blue planet has remained stable. So when rivers pour megatons of salts into the oceans how do these avoid acceleration into dead seas? How, also, has pH stability been crucially retained?

The Mother's bones, nails and hair are soils and solid, crustal rocks that, washed by storm and stream, yield minerals. These minerals life's producers, plants, absorb. Volcanoes throw up irregular formations like mountain chains whose various habitats permit an abundance of ecological niches. If life's sac is sky, the earth's skin is a crust of islands (or tectonic plates) that float on seas of magma.

Animate is coupled with inanimate. A system is a network of ideas or objects linked by common purpose. *By this definition life on earth is called an ecosystem.* An ecosystem includes living and non-living factors combined into a single self-regulating system. For James Lovelock's Gaia theory earth is a 'super-organism' made of all lives tightly coupled with air, oceans and surface rocks. Such a 'super-organism' maintains dynamic equilibrium. It comprises a totality that *seeks*, in a cybernetic manner, optimal conditions for life. Its variables include temperature, pH, salinity, electrical potentials etc. **'Cybernetic homeostasis', like 'program', is a conceptual phenomenon. Its presence in any machine indicates an underlying purpose.**

For their biological part individuals, populations or communities of different kinds of organism are involved. Life's geo-physiological health, the poise on which all ecosystems and their multicellular inhabitants depend, is in great part the gift of bacteria. Whatever their mode of origin, microbes of the kinds that exist today always existed. Tough and reliable, they toil relentlessly. They 'plough' the earth and continuously 'farm' organic substrates on which other organisms thrive. Bacteria might even, as the foundation of life's ecological pyramid, be construed as its primary, substantial, most important form of life; yet, working at the interface with inorganic matter, most 'robotic' too.

Provision, consumption, recycled waste. *You need all three.* Input, process, output. *Homeostasis needs all three.* Ecology is irreducibly, biochemically homeostatic and cyclical. Together every community and every level of life in each community cycles around each co-factor. Indeed, each organism plays one or more of the roles. You need three-in-one to peg the balance happily.

all below	*Transcendence*
lesser sorts of being	*Supreme Being*
balancing acts	*Equilibrium*
range of shadows	*Essential Light*
moral spectrum	*Good*
↓ *negative wish/ act*	*positive wish/ act* ↑
from Truth	*towards Truth*
descent/ darker	*ascent/ lighter*
body/ self-centred	*soul-centred*
passion	*detachment*
malevolence	*benevolence*
hate/ abuse	*love/ care*
crooked/ perverted	*straight/ open*
criminal/ demonic	*saintly*
a curse	*a blessing*
pain	*relief*
decline/ fall	*lift/ helping hand*
depriving	*sharing*
malefactor/ enemy	*benefactor*
evil/ sin/ immorality	*goodness/ virtue*
darkness	*lightness*

From positivity let's turn, before considering deliberate negativity, to **nature's shows of seeming negativity.** An ill wind, evil cold, cruel sea and other natural challenges (not least, inescapably, the body's own calamities) may threaten life with suffering. They may spell fearful pain and death. Such 'evil', as we understand, does not involve intent or animosity. It is, as matter is, oblivious: not immoral but amoral. Thus 'ill wind' does not blow with ill intent. It blows according to the fashion of inanimate design. In short, nature's so-called 'negativity' is really its insensible neutrality.

Voluntary negativity is quite another thing.[87] **Evil deals, intentionally, in lies and pain.** Its condemnation isolates, its burden weighs you down. Negativity inflicted on another is the state of hell.

Resistance, opposition and exhaustion - the world exists through shadows set against its light. Beneath transcendence truth is broken by polarity into opposing vectors; plus and minus are unbreakable a pair. But as free agents and not unconscious robots humans *choose* which way to act - and sometimes voluntarily create another's pain. Holistically-speaking, don't blame divine but human choice for evil. Thus, if evil's sourced in our own heads, perhaps moral struggle isn't cosmic but just local.

No-one denies that in a tough environment the bestial side of life is struggle for survival. 'Survivorship' and 'reproductive fitness' win a day that's governed by genetic products called your limbic system and its master glands. Furthermore, according to materialism's evolutionary logic, rapine, pillage, fraud and exploitation are thus implicit in your naturally selected genes.

What, therefore, might not a well-evolved and cunning despot's genes inflict upon his chosen enemies or, in communism's case, non-atheistic enemies of scientific reason? Purge, pulverise, imprison, forcibly re-educate? Eradicate all threats to his genetic egotism's article of faith - any-cost survival? Ruthless, forceful and, the best of all, efficient tyrants soon create such social 'excellence' as hell.

Who could not be, by selfishness and his desires, converted into tyranny? **Therefore, from passion, pain and isolation let us turn towards cure.** Down-to-earth solutions to the problems of disorder, ignorance and suffering involve **two species of utopia - one objective, outward social and the other inward, personal and subjective.** *We'll deal with the outward, large-scale solution first and the, work inwards towards a microcosmic, personal Nuclear Solution.*

'*External, outward association*' comprises relationships between an individual and his earthly circumstance. It includes both positive and negative interaction with non-living objects, any living organism and, closest, other humans. Regarding this let's simplify the situation from a Dialectic point of view - a **Unified Theory of Religion, Politics and Law. Religion, politics and law - this homeostatic triplex guides the individual and, by extension, his society. They serve the objective, outward social species of utopia and so their treatment is called <u>sociological.</u>** *Thus* *sociology is about your 'wider body' especially in the sense of interactions between individuals in human societies.*

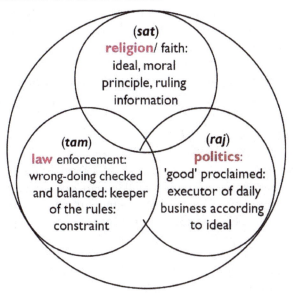

Here, derived from the cosmic fundamentals, is the social triplex. It is the previous diagram re-expressed.

body/ mind	*Soul*
lesser selves/ personae	*Self*
concentric rings	*Nucleus*
relative truths	*Truth*
relatively sane	*Sane*
↓ *external*	*internal* ↑
descent	*ascent*
outer exit	*inward return*
lower self	*higher self*
from Creator	*to Creator*
less sanity	*more sanity*

Religion (see Glossary) means 'a system of belief' and, whether vague or clearly defined, is unavoidable. Whether atheistic or theistic, oriental, occidental, holistic, humanistic or plain secularly 'liberal' it frames relationship with both cosmos and, in human terms, society.

For the various cults of materialism nature rules the day - by which is meant that non-conscious, amoral, undirected forces have produced mankind and his societies. Mindless matter has *per se* no purpose.

Everything, including cerebral mind, is made of matter; this non-conscious factor has no will; no will, no choice, no morality. If this is hopeless and a counsel of despair then, the creed exclaims, grow hard and strong because that's just the way this harsh, unfeeling, aimless and amoral cosmos is. At least, for the body's life-time, take heart from victories, intellectual pleasures, reproduction and, sensual survival with like-minded friends.

On the other, holistic hand, thought is not the same as things; and the quality of natural, universal mind is not the same as matter's. In a 3-step, projected cosmos mind is immaterial. An immaterial element exists and, in this case, the universe is *only part* non-conscious. **From this view derive two kinds of religion, one 'peripheral', one 'nuclear'.**

Mankind's 'peripheral' religions, the world's formal faiths, are social encrustations born of ritual surrounding nuclear experience (see Chapter 2: (*Sat*) Potential Information or Illumination). Saintly teaching is externalised in differing, local ways. Holy scriptures, creeds, imaginations, theologians, uniforms and prayer-houses represent an illumination-inexperienced congregation. This expression encourages good will and happiness; and at best it points (↑) upward towards a knowledge of the Peak Ideal. At worst, however, creeds breed isolation. Whence an oft-bellicose delusion of any particular organised religion is that it's 'the only way' or 'sole repository of truth'. Encrustation of ideals, sclerotic dogma, misinterpretation, superstition and misunderstanding lead to conflict and, destructively, to war. History is littered with (↓) perversion of the Nuclear Truth - The Natural Truth darkened, twisted round and downward into politics of fear, chastisement and control This egotistical perversion, personal or political, is what yields the poisons we call hell on earth.

degrees of alignment	*Ideal*
lower principles	*First Principle*
consequent development	*Nucleus/ Source*
belief	*Knowledge*
religious faith	*Supra-religious Fact*
↓ *darkness*	*light* ↑
peripheral formalities	*approach to nuclear core*
rigid adherence/ dogma	*thoughtfulness/ debate*
rule by fear	*rule by wisdom*
arrogance/ intolerance	*tolerance*
wrong	*right*

In these matters individuals on earth are not, except in soul, the same. Minds and bodies differ and, for these parts, **health** is the ideal. Body's health is given, mind's is chosen. Purpose, attitude and relative free will make choices. 'Good', 'bad', 'right' and 'wrong' exist; law and morality are real. In this case, *top-down*, paths towards ideal mental health always (↑) ascend towards the highest state of mind - love, compassion, happiness. They focus only on convergence towards the undifferentiated Soul-Point of an Absolute Morality.[88] So if, above mind, you treat with Nuclear First Cause, you treat with Metaphysical Ideal, Transcendence and Top Teleology; and with Free Will whose nature, Love, is paradoxically and by degrees constrained by various forms of mind. These conditional forms are, in this view, relative impurities of Ideal Quality.

Materialistic or holistic, either way *morality steps centre stage.*[89] Its values aren't of scientific kind but it is for sure the bigger player. The purpose of the agencies of social order - religion, education, politics and law - is to combat imperfections, minimise all criminal behaviour and promote the quality of relationships. *The whole point of morality is maintenance of individual and thence social balance and, wherever necessary, restoration of dynamic peacefulness.*

So which precedes - religion, politics or law? Is not the foremost regulator of the three, from which ideals derive, religious world-view? Secular or, if immaterial precedes and governs the material state of man, non-secular religion sets ideals. *Ideals, as far as realised, compose objective, social species of utopia.* In this case from Natural Dialectic's point of view best principle derives from Top Ideal. *Society's best formal answer is one that addresses evil and destructive issues from the State of Immateriality; our material solution is resolved, most definitely without recourse to science and without a test-tube anywhere in sight, by institutions charged with primary exercises in morality.*

Now from religion turn to *politics*, turn to the lower truths of egotistical pragmatics. Survey the action, business and economy of body - seething crowds, cross-currents of humanity; turbulence of social ocean, individuals jostling with each other in a way that keeps (or fails to keep) the peace. What is a crowd but a community, communities a town and towns writ large across a wedge of continent? Individuals in society, men in a collective state, states trading with each other - this is the bubbling cauldron, **human politics**, into which birth throws us and we boil. **As with biological and personal so with legal and political dynamics - resilient balance minimizes stress so that equilibration is, for body politic, the basis of its politics.**

Politics derive directly from peripheral religion whether its authority is secular (say, communistic) or not. Politicians govern and are governed by the influence of creed. Their truths derive from doctrine. Ideology will frame such laws as steer towards its quality of goals. Employing metaphysical ideal how best, philosophers debate, to legislate? Aligned with which criteria might a leader govern beneficially, cultivate dynamic equilibrium of decent living standards and keep the body politic in healthy peace? And also therefore, orderly behaviour in mind, of necessity construct sufficient fence of penalties in order to protect a subject's sense of equilibrium by maintenance of justice, law and order in a realm?

As well as being governed every human, following his or her adopted form of creed, self-regulates. Which, therefore, control or self-control, is best? Are conscience, self-control and an internal law to be preferred? Or don't they count and you invite a visit from the officers of external law? Rules leash animals with reason; they frame your practices with principle. You are constrained within a mental box of regulation. If you flout the ideals that it represents then an **external leash** of conscience must restrain. Those officers detain and teach you with the pain of punishment. They might even lock you in a hard box called a cell. This is the third (*tam*) part of a Unified Theory of Community.

States of mind. Chapter 2's key illustration shows that by involuntary senses we are drawn out and thus experience material forms; with feelings we respond to these and, specially, to forms of life; and also, with voluntary contemplation, we turn inwards towards our own Essential Being. In this pilgrimage towards Source we gradually transcend man's various systems of belief and are transformed. **Illumination is the goal of nuclear solution.** [90] **Its species of utopia is inward, personal and subjective.**[91]

Therefore, we now work inwards from Point X - our third eye (see Chapter 2 and Glossary). We focus towards the Real Deal, Nuclear Seed, Nature's Inward Positivity. No crime! No prisons, courts or punishments! Do such societies, outside monastic, still exist in which **self-government** eliminates the reckless, disrespectful element? Where, really, does self-government exist but mind? So crimelessness has great potential. Nature's Inward Positivity exists in every human but its exercise is voluntary and needs choice.

relativity	*Absolution*
duality	*Unity*
↓ *division*	*unification* ↑
material focus	*immaterial focus*
towards illusion	*towards truth*
no thanks/ ingratitude	*thanks/ gratefulness*

Delve further in. Take nuclear religion to your own (and everyone's) extreme. ***Internal, individual association* *involves just one relationship.*** How well do you live within yourself? How to realise The Real Deal - Your Self? And as far as possible return to Origin?

Nuclear Religion is about (↑) return from the periphery of creation to its Natural Centre. Isn't the Aspiration of a human life, naturally inlaid but much neglected and distracted, to distil the mind's pollution to a pure distillate and thus, as every mystic always told you, reunite your life with Life? Essential Psychological Unification: Communion: Return to Source lost, as the hymn sings, 'in wonder, love and praise'.

Thus, human evolution isn't one from man to ape.[92] True ascent is to discover your full measure, to realise your Full, Essential Potential and, thereby, the real, larger-than-Vitruvian span[93] of man's extent.

From the stack above, if Essence, Truth and Reality are Absolute and existence only relatively so, the Natural Dialectic offers a simple scale by which to measure and, against Criterion, prioritise these fundamental qualities. It would appear that existence involves a scale, more or less, of relative reality, truth, appearance or illusion - albeit (as both Einstein and the Saints agree) a persistent one.[94]

In conclusion I ask whether *information* may not be the hidden, immaterial factor that causes currently materialistic paradigms to shift. Whatever the case, I hope that you have enjoyed the exploration of a philosophical architecture whose *motif* is the binary pattern of Natural Dialectic. We have, considering both inanimate and animate, circled round the universe. *What is, from the evidence, the final line and highest of conclusions? What, at the most natural core of cosmos, is the nature of Truth? The choice, the faith is at the end between material chance and an Informant Creator.*

On a Broadcast Live from Mir (Peace)
Satellite to BBC, London, Christmas 1988

A peaceful eye surveys our ark
   from heaven. Unhastening sphere, the Earth
revolves through nothing, spins through wave
   on wave of life - womb, carrier, tomb,
our one and only hope and home.
   No boundary, no politics;

the sapphire waters, swirling clouds
   and land are granted us in trust.
Strip desire and burning as a candle-flame
   in 'Mir' treat Mother and her brood;
lead life as gently as you can,
   leave the planet spotless as you came.

Questions 6

1. What is the holistic view of 'absolute community'?

2. What is 'relative community'?

3. In the communities of Essence and existence what does X stand for?

4. What does your 'wider body' mean? Do you include it in your social calculations?

5. What three items compose the hierarchical triplex of community?

6. Arrange them as the triplex member of a 'stack'. Add a few more lines.

7. What facts, if any, persuade you that, whether accidentally or not, our planet 'ticks the box for life'?

8. What are the three basic co-populations on earth?

9. Can you name any decomposers?

10. Turning from health to pain, what is 'nature's negativity'?

11. What do we call the voluntary infliction of negativity?

12. What is the difference between an atheist and a theist?

13. What is the key 'social triplex' of institutions dedicated to the promotion of social and individual welfare?

14. Arrange government in triplex 'stack' form.

15. What is religion?

16. What is politics?

17. Why is morality 'difficult'?

18. What is the nuclear solution?

19. What is individual association?

20. Who, if anyone, is God?

Appendix 1: Simple Structure of Natural Dialectic

We pick up from after the oriental triplex near the end of Chapter 1.

↓ *tam*   *Sat*   *raj* ↑

Sat **(equilibrium),** *raj* **(stimulating ↑) and** *tam* **(inertialising or materialising tendency ↓).** We're almost there. *The first of three final steps* is to reframe these **triplex stacks as dualities. This is because the opposites that compose Dialectical stacks are a columnar expression of polarity,**[95] **that is, duality.** Let's start with a simple 'duplex' stack made up of vectored parts.

↓ *fall*	*rise* ↑
(-) *negative*	*positive* (+)
exhaustion	*stimulus*

A *stack* is a set (or pile) of members, in this case three members. Each *member* of a stack (e.g. fall *and* rise or negative *and* positive or) consists of a pair of polar 'anchor-points' or, simply, opposites. Each opposite (e.g. negative *or* positive) is termed an *element.* Each element is placed according its fundamental characteristic viz. its tendency to (*raj* ↑) rise or (↓ *tam*) fall. For example 'stimulus' is a placed in the right-hand column and its opposite, 'exhaustion' on the left. Health (right)/ illness (left), life/ death, happy/ sad - the world is full of opposites, the list is endless.

Of course, members of a stack are not synonymous. *But they are equivalent because their elements align in terms of character, that is, the tendency or vector direction of their cosmic fundamentals.* This can be called the theory's principle of equivalence.

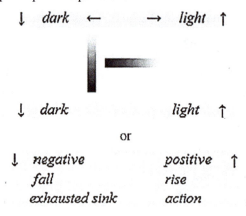

↓ *dark* ←	→ *light* ↑
↓ *dark*	*light* ↑

or

↓ *negative*	*positive* ↑
fall	*rise*
exhausted sink	*action*

Each member of a stack (e.g. dark and light) involves '**paired opposition**'. It implies a **scale or spectrum** (in the above example, one of greys) that runs between its elements; or, called '**complementary covalency**', it may imply degrees of attraction or repulsion. For example, 'more negative (↓), approaching extreme negativity' is set against more positive (↑), approaching extreme positivity'. Such scale is not numerical nor, if metaphysical members of stacks (e.g. purpose, thought, happiness and so on) are to be included, can it be. *Simply, as greys run between extreme black and white, so it is implied that a scale permits oscillation of values between any pair of elements.* It is asserted that, while a few isolated members mean little, the systematic, orderly linkage in stacks of particular interest can lead to eureka moments of discovery. 'I never thought of that before'. Such linkages can break conventional modes of thinking by establishing fresh relationships, insights and connections. Hopefully, you will become able to build your own stacks or 'connectivities'.

On any scale motion runs down (↓) and (↑) *vice versa*. **Thus a single, non-repetitious first-line representation of vectors, as on the previous stack, efficiently indicates the direction of elemental vectors for the whole stack.**

But what about the central (*Tao* or *Sat*) character? How is this treated in a binary system?

↓ (-) *negative   Neutral   positive*   (+) ↑

This triplex is resolved into two separate stacks. The first, top stack of the pair disposes Neutrality on the right against a neutralised description of the two polarities on the left.

↓ *polarity* ↑   *Neutrality*

The left-hand element is then split into its polar vectors, (↑) positive and (↓) negative, in a second, vectored kind of stack:

↓ *tam*	*raj* ↑
negative	*positive*
divisive	*unifying*
fixity	*flux*
lock-up	*freedom*

As was mentioned, for Natural Dialectic the ceaselessly moving, changing forms of creation are called existence. Their changeless Source is called Essence.[96]

↓ *existence* ↑   *Essence*

We can include this member, with or without vectoring arrows, in a slightly larger stack:

existence	*Essence*
relativity	*Absolution*
relative truth	*Truth*
polarity	*Neutrality*
duality	*Unity*

The equivalence of the extra four members offers a slight insight, which other stacks can build on, into the relative natures of Essence and existence. This kind of binary stack is called **Primary, Essential or Central Dialectic.** Having resolved a triplex stack into a binary, its central element (e.g. Neutral, Peak or Zone 1) is now indicated by writing in the right-hand, Essential column and using a capital letter.

Note that this Primary Stack sets (*Sat*) Neutrality against (↓ *tam*/ *raj* ↑) polarity; also Unity against (↓ *tam*/ *raj* ↑) duality and a range (or scale) of relativity against Absolution.

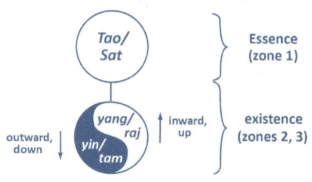

Summarising the cosmic fundamentals (where the Chinese and Indian labels are synonymous) and the narrative so far, this diagram also illustrates **the dependency of existence on Essential Source.**

Essence is Supreme Being without predicate. It *is*. It is Uncaused First Cause, the supra-existential root of creation. In Essence, forms appear; they 'stand out' which is what, from the Latin, existence literally means. Every force or form *is something.* It is a flat self-contraction (see Chapter 2: Aristotle) to claim (as, for example. did Stephen Hawking) that something - in his case gravity - created itself and then the universe.

From Unimaginable Void of Essence issues every object and event of changeful existence.[97] Thus the zone-numbers in this the diagram (and the following correspond to those in the model of Mount Universe.

Next, existence, whose duality implies polarity. **Such polar component is expressed in the lower, vectored so-called secondary, existential or peripheral/ polar dialectic.** For example:

↓ _tam_	_raj_ ↑
negative	_positive_
exclude	_include_
isolate/ specify	_generalise/ link_
materialisation	_dematerialisation_

Thus **secondary, existential stacks** (_written exclusively in lower case_) represent the various kinds of polarity from which the changeful web of existence is composed.

These two kinds of stack, Primary and secondary, can be combined to obtain a full picture, in any particular circumstance, of the three fundamentals. **This combination is called a linked or full stack.**

Primary, Essential or Central Dialectic:

tam/ raj	_Sat_
existence	_Essence_
polarity	_Neutrality_
expression	_Potential_
limitation	_Infinity_
duality	_Unity_
relativity	_Absolution_
motion	_Balance_
something	_Nothing_
zones 2 and 3	_zone 1_

secondary, existential or peripheral Dialectic:

↓ _tam_	_raj_ ↑
fall/ down	_rise/ up_
negative	_positive_
division/ multiplication	_unification_
isolation	_connection_
drag	_stimulus_
zone 3	_zone 2_

Here a **full** stack has been constructed by linking a binary pair. As will be shown, this can be done for any different context you may choose to clarify. In some cases, the logic may force us to change muddled preconceptions.

In the Primary part of this full stack all elements of the right-hand column reflect, by the principle of equivalence, the character of Essence. For example, Infinity, Unity and Nothing are not synonyms but, paradoxically, all represent aspects of Essence.[98] Of course, we build on identifying the character of Essence and of existential relativity throughout the book.

In fact, reading down any column of elements in any stack is an act of connectivity. Links, some obvious but others less so, appear within the pattern of any properly ordered stack of opposites.

A good, *top-down* way to read a stack is, as in these two diagrams, to start with the arrow above right of the Primary Dialectic's right-hand column; then track to its left-hand, polar column; from there zig-zag right to left with each member down the stack (e.g. Essence, existence; Neutrality, polarity etc.); then cut from base Primary to the active (*raj*) side of the secondary stack and zig-zag similarly down that stack.

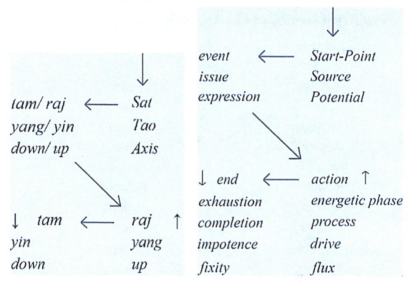

Start, action, end. From source (potential) through issue to exhausted sink. From start through process to completion. Every object or event involves prior precondition (called potential), an active phase and a passive, finished or fixed phase. **This, as we saw earlier, includes not only the course of all events, great or small, but the process of creation itself.**

In this way you can (amid myriad other words and definitions used by humankind) clearly define the pre-active source of cosmos, before any form has been created, as Motionless Potential.

Having **reframed triplex stacks as dualities** we now need, in the **second of three final steps,** to resolve a paradox viz. **the ubiquitous local reflections of Essential characteristics in motionful existence.** We need to discriminate between characters on the right-hand Essential stack and their finite appearances. For example, balance, unity and potential are 'zone

1' characteristics also occuring at myriad different times and places in zones 2 and 3 of existence. Source transcends mobile creation; it is **Cosmic First Cause but is, paradoxically, reflected endlessly in lesser psychological and physical causes. We distinguish, in effect, between Absolute Nature and its reflection in relativities, that is, in finite phenomena.**[99]

imperfect	*Perfect*
scale of lesser perfections	*Apex*
lesser beings	*Being*
lesser truths/ appearances	*Truth*

In short, the characteristics of the Essential column (capitalised, right-hand in Primary Dialectic) are all qualities of Supreme Being. Perfection is such a quality. It is commonly *reflected* as part of creation *but only in local, temporary forms*. Every existential object or event has boundaries; it is - psychological (in the subjective case of mind) or physical (in the objective case of bodies) - conditioned. All cases are constrained by form.

So you can logically build a picture of creation fthe Central One. Essential Perfection is unconstrained but generates a scale, a binary conscio-material gradient which, as chapter 2 demonstrates, iscomposed of proportions of informative and non-conscious, physical energies, you can understand the nature of a scale of finite beings called creation. We call such finite, boundaried and thereby relative realities '*lesser*', '*seeming*' or '*apparent*'. They are *phenomenal* (a Greek word meaning appearance) and therefore constrained in more or less degree. **Their lesser being, essence or reality is hierarchically arranged.** Nearer to Reality is, therefore, more real. What is closer to the Quality of Truth is more true. By this Criterion you gauge the nature of an act, event or thing.

Apart from Infinite Reality, physical time and space[100] are interesting *lesser or apparent infinities.* And, within these extents, myriad equilibria (called things) appear each with equalising, different weights on either side. These are *temporary points of balance,* changes we describe with our equations. All phenomena (including their neutralities, potentials, causes and so on) are relatively restricted transformations, that is, they are localised in time and space.

Next the *third step* - inversion.

It's apposite to define a digital phiolosophical structure such as Natural Dialectic in terms of 0 and 1. It reads 0 (zero or no action) in terms of both pre-active potential and its inversion, post-active sink; this 'switched-off' pair are opposites in character. Meanwhile, 1 (*a unit*) represents 'switched-on' action, that is, each energetic form of transformation. It thereby reperesents change or the transcience of all things.

123

existence	*Essence*
0/ 1	*0/ Off/ Inaction*
active expression	*Pre-active Potential*
issue	*Source*
↓ *inaction*	*action* ↑
sink/ exhaustion	*current*
post-active impotence	*energetic action*
0/ off	*1/ on*

In this stack note that 'inaction' occurs in both top-right and bottom-left columns. The word is the same but its nature completely different - as distinct as source from sink. Conflation of meaning here is just a verbal illusion. Similarly with the word 'peace' - potential (no action yet) and RIP (action over). Such spiral' down from pre-active to opposite, post-active condition is called **reflective asymmetry** or **inversion.**[101]

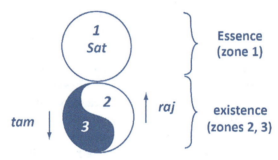

This picture simply reformulates the stack above using the 'yin-yang dependency' diagram you met a page or two back. A binary system intrinsically includes inversion between opposites or, if you like, passage from top to bottom, that is, Source (zone 1) to sink (zone 3) of a full stack. **It is the way creation works.** As regards cosmos (and, therefore, your microcosmic self) such inversions run from the expression of Informative Potential to its complete absence (or locked, impotent fixity) in non-conscious, automatic matter; and, as regards energy, a fall from subtle quantum elements to the gross, locked expression of solid matter. To repeat, two kinds of inaction - that of pre-active, poised potential and post-active, exhausted finish - are asymmetrical reflections of each other. They are, as are also the unchanging, general stability of precondition (in physics called law) and the changed, specific stability of a fixed outcome, entirely different. **These, informative and energetic, are the couple of *Primary Reflective Inversions* that, interlocked, compose the cosmic conscio-material gradient of creation or, if you like, the informative/ energetic scale of cosmos.** Such gradient is no figment; for example (as Chapters 2 and 5 illustrate) your own mental and physical structures reflect it.

Appendix 2: Equation/ Equilibrium

Between each side of an equation sits a connector, a central balancing factor in any transformation called an equals sign ($=$). Chemistry uses a single, one-way vector ($\rightarrow$) from reactant to completion, that is, product; also a two-way ($\rightleftharpoons$) 'equilibrium arrow' indicating that (re-)action can proceed reversibly. Points of equilibrium may vary along a scale. The one-way vector, turned vertical ($\downarrow$), also represents the way - from source to sink - that a full Natural Dialectical stack is read (*see* also Chapter 1); while the 'equilibrium arrow', so turned ($\uparrow\downarrow$), reflects both existential Dialectical vectors. Fulcrum or point of equilibrium is the Essential Dialectical non-vector which is also identified as Peak, Source, Potential and so on. This trinity ($\uparrow$ up, $\downarrow$ down and balance) constitute the cosmic fundamentals (Chapter 1).

Binary opposites are not, as Niels Bohr noted, so much contradictory as *complementary*; each is dependent on the other for its (and thus duality's) existence. As the yin-yang symbol (☯) illustrates, the seed of the other remains to some degree in each; but the extent of proportion may vary so that between poles exists a range of 'mix' (between energy and information), a spectral continuity of proportions. In this inclusive view opposites are each part of a greater, transcendent whole; they are dual components of relativity until neutralised in mergeance with the Absolute Balance of Unity ($=$ Essence, Source etc.; *see* esp. Chapters 1 and 2; also other books - *SAS* Chapter 8 and *PGND* Chapter 8). Although opposites may (positively) attract and (negatively) repel, the pair are observed in a continual attempt to equilibrate, to strike balance. Such equilibration, the nature of nature, is at the heart of transformations. Equation is in the middle of dynamic change and thus at the heart of its numerical description, mathematics.

Therefore, vectors, equilibration, equilibrium and the principle of equivalence (are also at the heart of our Theory of Complementary Opposites, Natural Dialectic (*see* Appendix 1 and Chapter 1; also Glossary for this quartet; and 'cycle') **In this view, myriad balances - local and transitory - forever subtend Absolute Equilibrium.**

In other words, local, changeable species of balance occur throughout creation and are called relative or lesser, existential equilibria. They amount to apparent, temporary fixations. **While transitory balance/ neutrality is the way of chemistry and physics, Absolute Balance is, on the other hand, the goal of psychological equilibration.** The Taoist/ Buddhist 'middle way' achieves zenith at its Centre-Point (called *Tao*); and throughout the orient the central cosmic fundamental, Essential Balance, is called *Sat* or *Sattwa* which is translated Truth. In fact all faiths, using different metaphors, strive to achieve this Natural, Metaphysical Absolution, the final compaction of detail into principle, duality into Original Unity.

Natural Dialectic admirably fulfils Einstein's aspiration of seeking the simplest possible scheme of thought that will bind together the observed facts. The search for such a scheme amounts to a Holy Grail (Chapter 2 and Glossary: Unification). And an equation, operating like a principle, represents a balance that compacts detail into simplicity. For example, the Mandelbrot equation (here autographed by Mandelbrot but lacking an iterative 'equilibrium arrow') generates a great variety of form; and Einstein's $E=Mc^2$ equates the two basic forms of non-conscious existence, mass and energy.

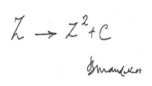  

A third equation sums up his Theory of Relativity (and thus how physical cosmos works) as 'elegantly' as possible. Such compressed description of physical archetype needs, as do compactions of holistic archetype like *Om* or Word, systematic unpacking to release its principles and detail. Interestingly, in this case the equals sign represents a channel of interplay between complementary immaterial and material components of the physical universe. On the left, the immaterial (or metaphysical) world of space, time, geometry and mathematics interacts with, on the right, the physical distribution of mass and energy. Space and time tell matter how to move while matter instructs space-time how to curve. Mathematics is certainly a metaphysical exercise. Chapter 4 therefore asks whether mind, maths or both together comprise the basis of our physical reality.

Process through the fundamentals, the way of reading every stack and describing every event, shows inversion, that is, reflective asymmetry

(Chapter 1). An example is the course of charge (potential) through current to discharge. *Symmetry*, on the other hand, is an aspect of balance, that is, of equilibration and equation. Various kinds of symmetry interest mathematicians, scientists and artists. Complementary equal-but-opposites e.g. matter and anti-matter or electrical (+ −) charge are ubiquitous; and the balance of complementary opposites, two halves of a single system, is common in nature. An information system involves sensation and response (Chapter 3); the bio-energetic system engages photosynthesis and its reverse, respiration (Chapter 5); and the reproductive system male and female parts. You are an example. And bilateral symmetry where the division of a whole is reflected along each side of an axis is also very common in biological forms. Nowhere is metaphysical symmetry clearer than the case of morality and, therefrom, law (Chapter 6).

A further quality of the (*sat*) characteristic of symmetrical balance and proportion is the derivation of aesthetic pleasure. Recall facial beauty or the elaborate counterpoint of, say, Sebastian Bach. Other wonderful examples of harmonic measure abound as, for example, in Mandelbrot and Tibetan mandalas or religious architecture and decoration the world over. The intrinsic aim of beauty is to promote the harmony of a balanced mind. More balanced is, automatically, more holy. This is the appreciation.

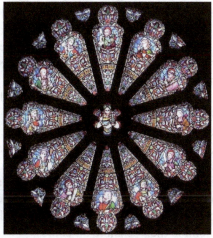

Appendix 3: Biological Archetype

Life's key aspect is exchange of information. A great deal of such exchange occurs in our bodies without us knowing about it. It is possible that, except in the case of some animals, no exchange of information is ever consummated in awareness. But this is not to deny any biological unit, including a 'dormant' plant, fungus or bacterium, natural and therefore intrinsic sub-consciousness. Each single or multi-celled type of organism with its appropriate natural protocols of information exchange senses the world, and responds to it, through the structure of its own chemistry and archetypal memory, that is, in its own typically programmed, automatic way. Materialistically, the whole game's made of molecules. However, holistically, if information is included in the mind=brain equation, then materialism's answer must be incomplete. We thought cells were simple but know now that our previous knowledge was (very) incomplete. To advance our present mode of thinking this Appendix follows up a neat, logical construction mooted in Chapter 3 (*see* Signal Translation, A Two-Way Psychosomatic Linkage). Check back to sections on the lower, unconscious subdivisions of the 'ziggurat', 'archetype' and 'grades of human bio-classification'. *The biological archetype is called a typical mnemone.*

Within this mnemone archetypal program is a magnet, attractor or prior organiser. Circumstances push but goal-oriented program also draws a process forward in time. Thus the real psychosomatic question asks precisely how connection is made between informative mind and forces, atoms and molecules of the phenotypic composition that confronts you in the mirror - your body. A simple magnetic field exerts influence; it organises iron filings round the magnet. A mind-field tries to organise the world to fit its own desires; and both conscious and sub-conscious patterns can exert direct influence over heart rate, breathing, body chemistry etc.

The idea that no metaphysical, informant influences exist is only odd to materialistic determination. But, if they do, why should immaterial symbol-carriers resemble their end-products. For example, 1-d *DNA* no more resembles 3-d proteins or bodies than electromagnetic broadcasts resemble the picture on your TV screen. Nor would we expect wireless constructions to resemble their sensible outcomes.

In this way it is reasonable, although the exact system flow-charts are unknown, to suggest that this memory is programmed, like a computer (also known as a mind machine), to achieve specific ends. Also, like a physical counterpart known as *DNA*, to suggest that its size, complexity, reasons and logic inhabit every cell of every body. It has been suggested that this subconscious databank comprises three cooperative subroutines. All inward traffic across the psychosomatic border (*PSI* or Synchromesh 2) passes from quantum level (patterns of charge and light) to a metaphysical **signal translation** routine. This modulates the input for issue to either or both of two subroutines called **instinct** and the **morphogene**. In the case of innervated organisms sensory data identified for translation into conscious experience is also passed through an extension, **an additional subroutine of signal translation** for passage across the interface called Synchromesh 1. And, of course, for outward motor traffic the reverse flow occurs.

Now, two different diagrams illustrate a *suggested* architecture of the subconscious with respect to conscious and unconscious (dormant) organisms. *Firstly,* the simpler program for *unconscious organisms* such as plants or fungi.

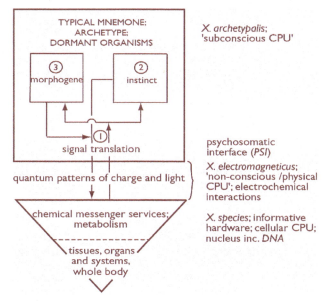

As illustrated, a typical mnemone is composed of *three* main subroutines; or, if a protocol is a standard procedure for regulating the transmission of data between two end-points, three linked protocols. **As noted, together code-translational, morphogenetic and instinctive programs comprise an individual's archetypal memory.**

Just as the genetic 'book of life' is found as a nuclear genome in every cell, so every cell accompanies its typical mnemone. Individual genetic material is passed from generation to generation of cell; the physical medium of transmission is either asexual division or a single fertilized egg. But the mnemone is not thus passed; you might liken it to a generic permanence that is automatically the metaphysical substance of every ephemeral cell. Destruction of dependent, local apparatus does not affect the independence of its file in archetypal memory. **Typical or generic pattern survives the death of any individual cell or body.** So archetype cannot become extinct.

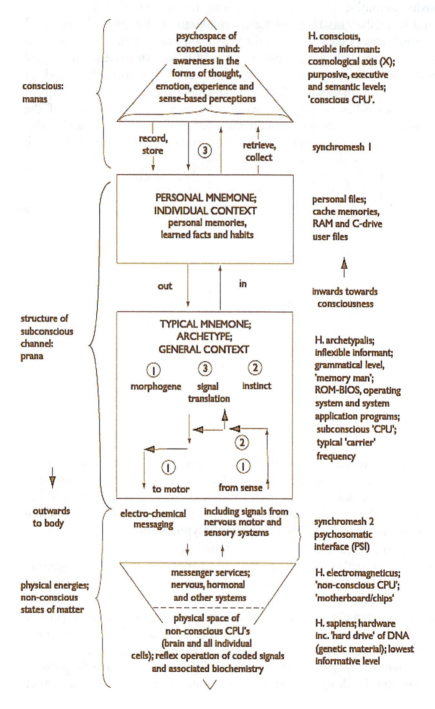

Can you, acting as a systems engineer, suggest improvements to the flowchart of a program 'computing' linkage of the physical world to unconscious and conscious aspects of mind and, since the traffic is two-way, vice versa? Lecture 4 of the One World Series leads to further exploration and examination of this important matter.

A mind-form, whether a transient thought or a filed memory, is as real as any physical object. Thus mind-objects, especially the fixed records of subconsciousness, are objects as real as anything physical - indeed, more so since they precede the latter. In this case, of what *is* first cause physical, archetypal memory, potential matter or typical mnemone (as the same structure is variously called) composed? **And, as working machinery, what is its fuel and what coordinated structures does it drive?**

Fuel that continuously supports its own creation has (although denied by materialism) been proposed as *Logos,* Word, *Brahman,* 'full range of nature' and many other names. One power, many names. This current, whose conscio-material gradient drops from psychological archetype to non-conscious phenomena, is the binary spectrum of existence. The relatively low frequency waveband that enlivens subconscious forms is called 'metaphysical electricity' or *prana.*

Mind grades from the clear light of super consciousness through the ever-active weather of conscious thoughts down to the 'solid' fixity of its databank called memory; and matter runs from particle field through gas to liquid to solid. It is therefore arguable that, in a holistic scheme of things, the lowest frequencies of psychological wireless, interface with the subtlest particles of matter - principally the electron with its stimulating charge. Such interface would form the psychosomatic border. But it needs be emphasized that in a holistic model of the cosmos each level is, as with Russian dolls, *nested* from Central , Source outward to the non-conscious, peripheral, physical level of our starry universe; and, since subconscious mind is the psychosomatic stage in vertical causation (Chapter 2), at its origin the entire physical universe must have been projected through this stage of universal mind. Also, unlike brains or any other atomic-based structure, mind is in all parts *wireless* and therefore what some might call 'non-local'. Other pertinent analogies include that of a power grid which transforms voltage in steps; the vibratory nature of sound and light; of a sliding guitar lick that moves smoothly though frequencies; distinct notes of musical scales; the integrated rhythms of harmony; and the informative power of integrated programs and their algorithms.

Within creation, energetic motions shape all events. Even 'hard matter' is, Einstein showed, composed of vibrant energy. **And waves of vibrant energy, in phase, show phenomena called positive interference and, in the case of a driving force, resonance.** Such associations increase amplitude. As any radio operator will say, attunement improves the quality of signal whereas negative interference creates cacophony or deletes signal altogether. If creation's current broadcasts at all frequencies then, in the psychosomatic case, the flowchart suggests a flow of subconscious *prana* will resonate/ positively interfere with the structures we call archetypal memories. Just as many frequencies are possible within the 'bracket' we call an e-m waveband, so millions of frequencies may be broadcast in what could be bracketed 'subconscious' zone. There is the possibility of very fine attunement that each mnemone and its various sub-files may, like an antenna, respond to.

Everything that cannot be sensed is not necessarily non-existent. **From holistic logic it can be inferred that metaphysical forms exist; we know that even personal memories can be complex (such as Homeric poems, the Koran or parts in plays all learnt by heart). They are indexed according to resonant association with other thoughts or with physical context such as a sight or scent. Thus we can reasonably infer that archetypal programs (each to its own organism) include complex networks of information.** As regards integrated complexity mankind has also discovered in the last century or so of human history the enormous complexity, full of signals, of what our bodies are made of, cells. Therefore, it is far from unreasonable to follow the logic and propose an extremely complex yet basically simple operation of networks or 'files' *within* networks of physical electrons. These configurations (and their wireless metaphysical, *pranic* counterparts) form the conduit of two-way psychosomatic traffic whose programs will be triggered by resonant associations.

We might also propose that each organism has its own typical *channel* from which are called subroutines. This channel is part of the *Logos* and so broadcast without cease. Whether a program or channel is picked depends on the presence of an antenna and its TV or, in this case, psychosomatic apparatus of an organism. Such a system is neat, efficient and coincides with not only cells but the morphogenetic shapes of phenotypes. Vitruvian man (Chapters 3 and 5) shows how subconscious energy is channelled through hubs of various frequencies called *chakras. Chakric* power-stations concentrate flow within an appropriate waveband of current that hierarchically relates to the triplex of types of body function - information in nervous system, energy regarding lungs and heart and the exhaustion of water and solid waste. Such systematic integration indicates that you are not a product of mindless, aimless matter.

Can this bundle of metaphors help us better conceive what's going on because, although holistically logical, it is not materialism's textbook hypothesis. And fresh ideas need concentration.

Do you remember vertical causation as witnessed in the simple finger waggle? Now we ask what instruments might be involved in the seamless, two-way psychosomatic transmission of information.

A mnemone is, we've learnt, a division of memory whether individual or universal. Typical mnemone is also a synonym for natural, archetypal memory in universal mind; it is in this respect part of a body's metaphysical *DNA*. And, as a computer program may be divided into interconnected sections, so is the mnemone. Its triplex division consists of cooperating subroutines - *signal translation, instinct* and *morphogene. It is to a study of the triplex nature of these programs that we now turn.*

Morphogene

It has been argued that, in a holistic scheme of things, the lowest frequencies of psychological prana interface with the subtlest particles of matter - principally the electron with its stimulating charge. Such interface would form the psychosomatic border. *But it needs be emphasized that in a holistic model of the cosmos each level is, as with Russian dolls, nested from Central Source outward to the non-conscious, peripheral, physical level of our starry universe.* And, also, unlike brains or any other atomic-based structure, mind is in all parts *wireless* and therefore what some call a 'non-local' field of influence.

What exactly is an electron or charge? Humans rely on huge electrical supplies but no-one has seen an electron nor understands what it is made of. Holistically, it involves a unique metaphysical vibration. In this there is no defference between mental (*pranic*) and material electronic charge. **They are, at different 'octaves', the same thing. Therefore there is no need for antennae, transmitter or receiver because they are, very efficiently, different levels of the the same. In this sense you could call the psychosomatic or mind-matter junction a seamless transmission zone.**

When an impression is made on the body it is translated into nervous impulses (which are still physical) through to metaphysical mind. The latter may or may not contain a file of electron congurations. Remembering the association of vibration/ frequency to form (Chapter 3) these subconscious holographs may naturally store shape. Such morphogenesis need not be confined to a single electronic network but could include those required for various parts that, in turn, fit the symbolic forms of developmental and adult shapes. Phenotype is thus indexed as regards its parts. There is a natural concept of type (such as human or oyster). While the exterior, physical environment may wreak changes. The typical concept remains unchanged. It is this conceptual ideal, towards which a body always cleaves, that we call *vis medicatrix*. It is both origin and reference point for an organism's 'ideal health' in body and mind, a homeostatic norm against which a constantly changing environment (both of the body and outside it) is referenced. Think of the morphogene as a system's 'signalman' that triggers switches; or like *CAD* software that an engineer employs, an image-source able to project and respond to 3-d shape instructions. Though still relatively crudely, such engineers use not pixels but 'voxels' (3-d datapoints containing multiple instructions for colour, texture, material and so on) to 'print' planned objects from a 3-d printer or scan, using f*MRI* machines, specific activity throughout the brain. Could nature have again preceded man, this time producing holographic 'voxellated' bodies? **At this point recall morphogene is the 'cache' of structural order; and that molecules and bodies are small and large-scale electronic configurations.** It is an organising template for electronic configurations of small-scale molecules, nervous networks of impressions and full-scale bodies. A 3-d configurative fingerprint correlates with its *pranic* self and if, as a databank or library, the morphogene carries a file for every biochemical and body part, then it may automatically 'recognize' what a given cell that works to build or sustain a needs or needs to expel. In this respect an 'undercurrent' of information is working in concert with codified programs whose metabolism generates target biochemicals.

An individual's two divisions of record are *personal mnemone* (likened to a short-term *RAM* data store) and *typical mnemone-*(likened to a *ROM* or an operating system). As 'background' and unchanging rules govern the flow of information and its various possibilities, so morphogene acts at the psychosomatic door as the guardian of orderly behaviour. We might place *personal memory* as a separate database interconnected with rest of typical mnemone.

Any organising template exists independent of the content that it frames. A machine is made of parts but coherence, integrity and purpose are its frame. *The whole is greater than the sum of its parts.* And, as Chapters 5 and 6 show, that greater part is mind. Similarly, chemicals and their interactions constitute the divisible parts of an organism but they do not sum to its structural and functional whole - the coherence, integrity and purpose reflected in its shape. The role of a morphogene is the unification, coherence and thereby governance, in terms of resonance with electrical patterns, of a life-unit, an organism. Biological *H. sapiens* is, as well as the phenotype of a genotype, the gross expression of electrical interactions that constitute *H. electromagneticus*. Each factor is physical, inanimate and essentially insensitive. Their reflex patterns are a gross reflection of an independent organiser - the subtle, sub-conscious, metaphysical infrastructure called *H. archetypalis*.

A morphogene, the generic form of storage for a biological program, is an organism's shape and working-dynamic in principle; and genetic practice supplies the detailed, individual differences of phenotype-on-theme. Just as mind responds to sense perceptions from its environment so a cell responds, by way of electrodynamic signals, with its physico-chemical context. The archetypal morphogene operates like a 'control satellite' off which specific, local *DNA* and other molecular signals are bounced; it is the 'brain' whose subconscious map coordinates those references and thereby integrates the operational hardware of a unit called either cell or, cohesively, multicellular body. Your own 'brain', replete with sub-routines, orchestrates *H. sapiens* as a conceptual whole through the medium of *H. electromagneticus*. It therefore includes a developmental sequence with, as a final 'still', the adult form.

Instinct

Auto-pilot. Default mode. This subconscious faculty involves routine - *instinct*. Instinct is a term that covers, in broad terms, reflex actions that are neither taught nor acquired. Such 'auto-pilot' is always purposive in scope and involves behavioural strategies of varying complexity. Purpose, strategy and tactic *anticipate* an outcome. Whether it is timely switching that supplies a metabolic pathway, an organ's cyclic function or whole body's integrative purpose to 'live on', all ingredients of life involve, as well as the requisite biological tools, conceptual know-how and, therefore, mind-behind. *The issue is not only operation but also the origin of programs of behaviour, that is, of information.*

Bottom-up genes make and modify all instinct. Genes alone make a spider spin its web, generate complex courtship rituals and, at root, will even let you think! Maybe, however, you don't buy half-truths. Maybe thought and instinct aren't just functions of your *DNA* and a magnificent but as-yet unexplained development, the brain; maybe that, *top-down*, there's a rational, prior place for immaterial mind. Cell, organ and organism are like aerials that resonate with information. And, if archetypal resonance rings possible, why not two or three mnemonic channels? **DNA sequences are already known as a chemical data storage system that is incredibly responsive, specific and precise. Could not an immediate, radio system with its conceptual broadcast, archetype, be prior, proactive and equally precise?** Doesn't an essential link, mind-body psychosomasis, engage communication of a subtle, rapid, highly-programmed kind?

A *top-down* point already made needs hammering home. Instinct is innate and unlearned ingenuity; it is problem-solving often in sophisticated ways. Spiders' networks shout the world is not material alone. As well as web-life's tricks instinctual memories include call-signs, food-finding, hunting strategies, migratory navigation and complex, distinct breeding patterns such as courtship, home-building, nursing and rearing; and, when life bodes ill, what about a creature's instinct for effective natural medicines (zoopharmacognosy)? Instincts are crucial and ubiquitous. **Their evolution is, of course, a Darwinian black box because their aspirational exercise isn't physical alone.** Indeed, at root, they're immaterial. Metaphysical. You need a framework, a broader box that's outside but inclusive of the biochemistry. Within this box the instinctive aspect of mnemone works as an ancient, indelible and often complex inscription. **It is a psychological datum,**

immaterial software, an 'under-writing' of intelligence, a common theme whose various strands are woven in the warp of archetypal memory. Auto-pilot is, therefore, an aspect of a universal mind.

Of course, any program's expedition involves physical kit. After all, the life's on earth. In the case of instinct genetic, hormonal and other apparatus trip the 'job' - but is job derived from them or do they serve its plan? If so, whither did this plan, a pattern of behaviour, evolve? There is nowhere evidence that an instinct evolves through stages of inferior skill or inadequate function. **Natural Dialectic therefore proposes that instinct, because it is the cause of such an obvious effect as behavioural goals, is our closest sensation of the immanent, numinous presence of archetypal memory.** In the case of gradual neo-Darwinian evolution, on the other hand, instinct and associated physiologies are, *as if yet another black box*, incoherently explained.

Is there any part of biology that does not exhibit breath-taking engineering, design and innovation in order to achieve specific ends? Nature (although perhaps not wholly material) is, in fact, a consummate chemist, physicist and engineer. Well-educated men, armed with chemical technologies, have now come lately to the natural table and demonstrate appreciation with an exercise dubbed '*biomimicry*'. Where nature got their first, they take inspiration from the codified ideas behind ubiquitous bio-technologies. Who in history suspected that every cell of every body was codified? And that translation mechanisms were in place to transform a passive, 1-d chemical expression of code into the 3-d shapes of an automated factory; and all the other shapes that compose myriad intricate devices and efficient mechanisms every body's system as a whole employs? Whence, though, logically do *code* and *program* come from? Mind or chance? Chance, guess Nobel laureates. And yet, since mind's action cannot be directly observed, the notion of morphogenic, archetypal code has been ignored. So Natural Dialectic forces holistically logical conclusions and suggest some answers that these pages have already started to propose.

Signal Translation

At the psychosomatic border *all* signal translation passes through the subconscious morphogene, instinct and personal mnemone. **However sensory input may mark a signal for transformation beyond typical and personal mnemones to the frequencies of conscious mind, that is, the realm of awareness.** This additional subroutine passes without reference to matter across an interface called Synchromesh 1. And, of course, for outward motor traffic from wilful minds the reverse flow occurs.

Of course, we know that translation from nervous impulse to conscious experience occurs but experimental neuroscience has not considered a hierarchical process that involves metaphysical mechanisms. We do not know how but the process certainly involves governance, that is, a codified program of physical and metaphysical cooperation. It therefore demands corresponding structural and functional unity. *Only in this way can the coherent propagation of hierarchically organised signals occur.*

How, then, to sum this Appendix up?

First to say, its contents are elaborated in The One World Series Lecture 4).

Bottom-up, developmental/ morphogenetic information is construed in reductionist terms as the product of chemical activity accidentally coordinated piecemeal, aimlessly, over aeons.[102] No immaterial factor exists.

Top-down, the basis of biology is information. If it is, like the proverbial elephant in the room or factor hidden in plain sight, then a materialistic vision of information-generation is deficient. A *hierarchical* structure of control[103] (see Chapter 3: *H. archetypalis* in biology) would suggest a step *up* from the molecular structure which creates an 'aerial' for controls at a metaphysical level. Such a level is named in humans, for convenience, *H. subconscious*. We have glanced at a possible structure of three main conduits of information-flow. Like the parts of a machine these integrated parts of memory act involve programs and files, possibly by the million, which regulate psychosomatic transmission and thereby the minds's seamless interaction with body and *vice versa*. Archetypal memory, also known (*see* chapter 2) as first cause physical or potential matter, is a crucial phase of *prakriti* (*see* Glossary) and of every cell of biology.

Glossary

adaptive potential: involves pre-programmed, super-coded switches and recombinant (transposable) refinements intrinsic in the genomic program of any particular biological type (*SAS* Chapter 23); the idea of such front-loaded potential for variation in the form of regulatory flexibility, laughable to generations of mutant/ junk *DNA* scientists up to the 1990's, may be confirmed by the presence of a nexus of nearly a million regulatory elements so far discovered in the human genome by the Encode Project; regulatory systems indicate intelligent forethought; they are always mindful with reasons, that is, with rationality; and are coordinated and incorporate high levels of systematic programming; in this natural case such density of information, packed in *DNA* molecules of which 100's of 1000's could fit on a pinhead, is even less likely than IT computer chips to arise by chance.

ahamkar: pre-scientific term; conscio-material band/ grade - conscious band of mind; sometimes identified (only partially correctly) with *ego*; faculty of self; *ahamkar*, involving identity, frames thought; habitual identification with own physical body; also with friends, family, community, study, work, country or cosmos as a whole; intellectual analysis is a 'knife' that dissects according to the interpretations of this frame.

anti-entropy: *see* **negentropy.**

archetype: basic plan, informative element; conceptual template; pattern in principle; instrument of fundamental 'note' or primordial shape; causative information in nature; 'law of form'; nature's script; Natural Dialectic's 'holographic' edge; the psychosomatic place where metaphysic and its physic meet; morphological attractor or field of influence in universal mind; the subconscious component of universal (natural) mind comprising archetypes; prototype-in-mind (maybe related to Platonic ideas, Aristotelian entelechies and/ or Jungian archetypes) whose potential matter is seen as hard a metaphysical reality as, say, particles are physical realities; like mind, being metaphysical, archetypes are physically non-existent; they are unobservable except by inference; materially formless and boundless, they are in this sense infinite; the fixed source of projection, an unconscious archetype, is omnipresent and omnipotent; conscious Archetype (First Cause Psychological) is active (not a memory) and also omniscient; as all orderly processes involve a program, so the passive, natural preconditions for our material universe are stored in cosmic memory - simple in terms of inanimate physical 'law' (of particles and forces), complex in terms of animate structure/ function/ behaviour; archetypal information is stored in a typical mnemone; in biology, this mnemone is the metaphysical correlate of biological type/ super-species; it amounts to the *potential*, that is, blueprint or codified meaning behind physically expressed form; abstract or metaphysical precursor; the collective unconscious of a type e.g. human type; as thought is father to the deed or plan is prior to ordered action, so archetypes *precede*, hierarchically and temporarily, physical phenomena; pre-physical initial condition of matter; check Glossary *tanmatra*.

Archetype*:*   Primary First Cause; *Logos,* material projection of conscious light giving rise, in the forms of its shadows, to a conscio-material spectrum (*see* also Index).

ATP: Adenosine TriPhosphate, life's standard bearer of chemical/ heat energy; a cell's agent of energy transmission; a biological 'match' or 'battery'; an active cell may discharge many thousand units of *ATP* per second to drive its metabolic machinery; the agent of recharge is a codified nano-machine, a 31-part molecular complex called *ATP* synthase; numerous in every cell, this rotary, respiratory dynamo is constructed of precisely sized, shaped and integrated components and, idling at 6000 revs per minute, its wheel-like turbine grinds out three units of bio-potential (*ATP*) per revolution; trillions of these molecules create your own weight of recharged *ATP* per day; you would, if they failed, be dead before you dropped; as well as a central, high–fidelity performance in the energy department *ATP*ases also play a critical informative role in the transmission of nervous and possibly other signals.

B

big bang: *see* **transcendent projection**

black box: process or system whose workings are unknown.

buddhi: pre-scientific term; conscio-material band/ grade - conscious band of mind; faculty of intellect; instrument, whether sharp or blunt in an individual case, of learning and discovery; analytical tool to educe physical and metaphysical patterns; pragmatic and hypothetical power of reason; crucial to gauge physical circumstance for survival and metaphysical principle for optimising state of mind.

C

caduceus: staff of Hermes/ Mercury the communicator, intercessor and informant deity; the messenger of metaphysic, carrier of thought is a power mythologically trivialised; symbol, including double helix, used by Natural Dialectic to represent basic human infrastructure, that is, the archetypal form of man.

chakra: pre-scientific term; conscio-material band/ grade - subconscious mind; metaphysical modulator; a distributor of different frequencies of subconcscious energy (*see* universal man chapters 3 and 5 and Glossary: *prana)*); *each chakra* is a hub that, like a sun or stars, influences events within its region; in the human its node of subtle anatomy is located close to neurological plexuses, suggesting close association of metaphysical with physical interaction; *chakras* are likened to power hubs like suns, 'controllers of regions') which exist on a cosmic as well as individual-body scale; they are correlated with *tattwas, tanmatras* (see Glossary) and cosmic 'elements' or states of matter today called solid, liquid, gaseous, energetic (inc. heat and radiation) and etheric (variously known as 'upper air', space and psychosomatic, metaphysial potential matter); *chakras* are a mnemonic device for the wireless transmission of *prana* to the electrical systems (e.g. cellular, nervous) of an organism also possible correlation with acupuncture points and meridians; they are *psychosomatic* gates; and thus a transit-point for informative signal, that is, trans-dimensional traffic (metaphysical to physical and *vice versa*); their mechanism includes antenna (receiver), transformer (between 'voltages' of *pranic* energy, transducer (between electrical and *pranic* conveyance of charge), simple harmonic oscillator (a 'heart' controlling *pranic* flow) and distributor (of *prana* through a network of meridians); they are archetypal channels, that is, specific mind-body broadcasting interfaces and the lowest metaphysical component in the hierarchical transmission of power throughout macrocosmic creation and its microcosmic reflection as mankind and other forms of life on earth; psychological chakras (or focal concentrations) are not included in this physical cosmology; apparatus such as copper electrodes, electromyographs and photoelectric cells have been used to help with scientific observation and experiment with *chakras*.

chaos: a confusing notion with three main but disparate implications - emptiness, disorder and randomness; Greek word meaning chasm, emptiness or space; structureless 'profundity' that pre-existed cosmos; *prima materia*, or primordial energy structured by regulation of divinity archetype or natural law; anti-principle of cosmos i.e. disorder; any case of actually or apparently random distribution or unpredictable behaviour; also apparently random but deterministic behaviour of systems (e.g. weather, electrical circuits or fluid dynamics) sensitive to initial conditions.

chemical evolution: also called abiogenesis, biopoesis, chemogenesis or prebiosis; implies that lifeless chemicals 'evolved' to the point whence they could 'self-construct' the primary unit of life, a reproductive cell; it means the generation, perhaps gradually over a long period of time, of life from non-living components by physical means alone. This process is integrally part of, strictly not the same as, Darwin's consequent evolution.

chitta: pre-scientific term; conscio-material band/ grade - conscious mind; attention *per se*; pure, formless (or boundless) intelligence in which forms of thought are projected; source of ideas and creativity; psychological focus.

chloroplast: organelle in plant cells containing photosynthetic apparatus.

chromosome: a 'book' in the 'encyclopaedia' of life; the human genome contains 46 chromosomes.

code: the systematic arrangement of symbols to communicate a meaning; code always involves agreed elements of morphology (the form its symbols take), syntax (rules of arrangement) and semantics (meaning/ significance); without exception such prior agreement between sender (creator/ transmitter) and recipient involves intelligence; given such reality a common disconnect, due to theory-driven, abductive speculation, is that natural forces might (as in the case of biological organisms) 'accidentally' generate code; in fact, nowhere in the universe is physical nature observed to generate objects of codified purpose or act teleologically; denial of this fact is a delusion.

Communion: the Christian term for mystic union (see also transcendence, *samadhi, nirvana*, holy grail); realisation of Universal Truth; Salvation; Absolution; core subjective experience; attainment of Supreme Being; Core Psychological Experience.

conceptual biology: holistic perspective allowing that, as in engineering or computation, metapthysical idea precedes physical expression of all biological systems and appliances; two basic principles are informative

140

archetype/ metaphysical concept/ plan and, from the discipline of information technology (IT), *top-down* programming/ code; this interpretation of biological and palaeontological facts should, *as standard*, be compared with the materialistic, evolutionary narrative; also called 'computational biology' or 'intelligent bio-design'.

conscio-material spectrum: also called conscio-material gradient; illustrates the basic components of polar existence; complementary informative and energetic components are thought of in a sliding scale of proportion; this scale drops from Source (Psycho-Logical, Informative First Cause) through grades of mind to non-conscious sink (non-conscious matter); creation thought of as a projection of light with accompanying proportions of shade - a cosmic chiascuro of every possible tone; source of this spectral projection is termed First Cause (Archetypal Potential) and is understood to be a Concentrate or Singularity of consciousness; the secondary first cause originates at the band of zero consciousness, 100% non-conscious energy; this concentrated singularity materialises; thus cosmos is viewed as the gradual embodiment of Uncreated Source; it represents a scale of possibilities expressed as typical yet individual forms; by this token an embodied particle of Uncreated Source (Soul) is subject to individual incorporations (psychological and physical); these relatively dynamic forms constitute its psychological and, in the biological case, bodily circumstance; Natural Dialectic also simply models such a hierarchical description of a di-polar creation by the use of concentric rings, step-wise, ziggurats and the orderly, harmonic arrangement of frequencies (music).

cosmic fundamentals: cosmic psychological and physical qualities; three basic states or tendencies labelled, for convenience, *Sat, raj* and *tam* or *Tao, yang, yin*; universal ingredients whose mixture is variously expressed in every object and event.

cosmological axis: human pivot; the point at which subjective and objective perception meet; eye-centre; third eye; thought centre; *ajna chakra.*

cosmological principle: idea that, on a sufficiently large scale, the distribution of matter in the physical universe looks just about the same from any vantage point; it therefore has neither centre nor, being infinite, edge - unless of course, its space is somehow spherical.

cosmos: often applied to physical universe, universal body; from Greek word meaning orderly as opposed to chaotic process; involuntary pattern of nature; also equated, including metaphysical mind, with existence as a whole; seen, dialectically, as a projection through the template of metaphysical archetypes; **the umbrella title of the series and website of books, Cosmic Connections, could, with reference to the Natural Dialectic which structures its *CUT* (see Glossary: unification), equally be called Orderly Linkages**.

creation: origination; physical or psychological arrangement; mind creates with purpose, matter without; creation means active production but also passive result; a creation will have been informed by force of mind and/or matter.

cycle: cycles cage nature - sometimes many kinds of them in what we call a single thing; for example, from a menstrual cycle your life cycle was developed; this cycle is, in turn, based upon atomic oscillations (atoms) and dynamic balancing (homeostasis) that lend to life's diurnal cycles bodily stability; not only you but, borne by cycles, suns, galaxies and planets swing around and round; motions round fixed axes spin as circles or, if the centre-point extends through space as well as time, wave-forms appear; waves carry energy, their repetitious frequencies define dynamic equilibrium; to-fro vibration round or through a 'zero-point' or 'norm' is what shapes cosmic regularities; and such contra-operation of binary vectors (e.g. ↑ up, down ↓) is central to the working of Natural Dialectic; perhaps the finest pattern of symmetrical polarity is light whose spectrum is likened to the conscio-material spectrum of creation in its entirety.

D

dialectic: a form of debate between positions of polar opposition (argument and counter-argument or thesis and antithesis); the motion of to-fro discussion that results in resolution (synthesis) whereby points of view are aligned; balance, compromise, neutral ground, golden mean and central truth are aspects of this synthetic (*Sat*) fundamental; paradoxically, two become one; union supersedes division; Natural Dialectic suggests that dialectical motion reflects the binary, cyclical nature of cosmos; such polar, to-fro or oscillatory dynamic occurs as the continual disequilibrium of nature (called motion and transformation) always seeks its various re-balances (see also Natural Dialectic).

dialectical stack: stack of complementary opposites; columnar expression of polarity; digital structure; a binary (base 2) *philosophical* system rather than (as in the way of Liebniz, Boole, Turing and computers) mathematical expression; there are two kinds of stack - primary or non-vectored and secondary, vectored; **primary (essential) stacks** set (*Sat*) Unity against (↓ *tam*/ *raj* ↑) duality (for elaboration see especially Chapters 1, 2 and Appendix 1); **secondary (existential) stacks** represent the various kinds of polarity from which the changeful web of existence is composed; each pair of polar 'anchor-points' implies a scale or dynamic range that runs between 'paired opposition' or 'complementary covalency'; stacks do not necessarily list synonyms or make equations; *their perusal is intended to promote connections because consideration of connections/ relationships tends to help unify/ collate/ organise one's working comprehension of any matter in hand.*

* Be clear, therefore, that the ordering of elements in members is always schematic according to the triplex operation of fundamentals - **from pre-active through active to post-active phase of any process.** Such universal process is self-obvious, simple but also the starting-point for an organised pattern of many connections in many contexts. Things link up in ways perhaps we hadn't grasped before.

* In this book stacks are arranged as dialectical points of reference relevant to a particular context. Examples are 'Primary and Secondary Stacks', 'Three-tiered (triplex) Mount Universe, 'Holy Grails', 'triplex mind', 'energetic cause of life on earth' and so on.

* A student first attempting to use the grammar of a fresh language inevitably makes mistakes. In the case of s student first trying to generate their own stacks the absence of a dedicated primer, tests and teacher

142

makes things trickier. (S)he should use the book's templates, based on the fundamental *sat–raj–tam* triplex, to help with alignment and choice of opposites. As ever, practice promotes fluency.

DNA: a complex chemical; a large bio-molecule made of smaller units, nucleotides, strung together in a row; a polymer in the form of a double-stranded helix; a medium superbly suited to the storage and replication of 'the book of life'; 'paper and ink' on which the genetic code is inscribed; an organism's 'hard drive'; *DNA* is such an elegant, efficient and densely-packing form of information storage and manipulation that *DNA* computing by humans is now a fledgling technology; in this fast developing field silicon-based technologies are replaced using *DNA* and other bio-molecular hardware.

E

electromagnetism: physics of the field that exerts an electromagnetic force on all charged particles and is in turn affected by such particles: light/ e-m radiation is an oscillatory disturbance (or wave) propagated through this field; light; light paradoxically involves a perfect, polar balance between contractive/ magnetic and radiant/ electric components.

elementary particles: science has discovered and, for the most part, experimentally verified, over fifty elementary particles; these are divided, in simple terms, into bosons (force carrying particles) and fermions (separate particles); bosons include photons (which mediate the electromagnetic force), gluons (which mediate the strong nuclear force), W and Z particles (which mediate the weak nuclear force), possibly gravitons (which mediate the gravitational force) and also possibly a Higgs boson (which may mediate a proposed mass-giving field); fermions include two main groups - six quarks and leptons (six electron/ neutrino types); derived from quarks are strongly interactive composites called hadrons; hadrons include baryons such as protons and neutrons and (perhaps a little confusingly) bosons such as short-lived mesons.

entropy: a measure of the amount of energy unavailable for work or degree of configurative disorder in a physical system (see second law of thermodynamics); inertial aspect of an energetic, material or conscious gradient; diffusion or concentration gradient outward from source to sink; drop towards 'most probable' outcome i.e. inertial slack; a measure of disintegration or randomness; expression of the (*tam* ↓) downward cosmic fundamental; a major property of matter, closely coupled with materialisation; in a closed system, which the universe may or may not be, this tends the eventual loss of all available energy, maximum disorder and the exhaustion of so-called 'heat death'.

enzyme: protein catalyst without whose type metabolism (and therefore biological life) could not happen.

epigeny: genetic super-coding; contextual punctuation; chemical modification of *DNA*; also extra-nuclear factors that may cross-reference with genetic expression.

equilibrium: three modes of equilibrium are (*sat*) balance of poise or pre-active potential; (*raj*) dynamic balance occurring in all regular cycles, wave-forms and cybernetic homeostasis that is basic to the stability of life-forms; and (*tam*) inertial equilibrium that results from diffusion of information or energy; it equates with exhausted inaction or 'flat', impotent rest; such post-active inertia represents the most probable distribution of energy/ matter with the least energy available for work viz. the most random arrangement permitted by the constraints of a system; expressed in psychological terms as ignorance, unconsciousness or sleep; see also equilibration, *karma*.

Essence: (*Sat*) Supreme or Infinite Being; Substance (perhaps Spinoza's Substance) 'prior to' or 'above' existence; Pure Consciousness/ Life; Peace that transcends all psychological and physical action; the root of an essentially undivided universe; Uncreated One within which and whence all differences have their being; Apex of Mount Universe; goal of saints/ 'philosopher kings'; the 'point' at which All-Is-One.

eukaryote non-prokaryote; any organism except archaea, bacteria and blue-green algae.

evolution: there are today *four* main usages of this word; each 'loading' derives from the original Latin, 'evolvere', meaning to unroll, disentangle or disclose; the *first two*, physical and biological, are conceived as natural/ mindless processes; the *second, mindful pair* is of psychological/ teleological import; specious ambiguity may conflate or switch between the fundamentally separate pairs of meaning. *Firstly*, in the scientific context of physics and chemistry, the word is used to describe process *not* according to codified instruction; it describes change occurring in physical systems; the laws of nature can't, it seems, evolve through time but stars, fires, rocks or gases can. *Secondly*, though also subject to the 'rules' of entropy, biological evolution is a theory of *random progression* from simple to complex bio-form; it thereby implies increasing *codified* complexity; while retaining the 'hard loading' of physical science it also, ambiguously, claims that codes, programs, mechanisms and coherent, purposive systems - normally the province of mental concept - self-organise by, essentially, chance; such confusion, the basis of naturalism, is compounded by failure to distinguish between, on the one hand, ubiquitously observed variation (called micro-evolution) and, on the other, Darwinian 'transformation' between different sets of body plan, physiological routines and associated types of organism - such 'black-box macro-evolution' as is never indisputably observed; to evoke a naturalistic ambience it is fashionable to use 'evolved' interchangeably with or to replace the words 'was created', 'was planned' or 'designed'; finally, it is noted that the coded, choreographed development of a zygote, packed with anticipatory information, through precise algorithms to adult form is the absolute antithesis of blind Darwinian evolution. *Thirdly*, man certainly evolves ideas; intellect can evolve 'purposive complexity'; we invent all kinds of codes, schemes and machines; we devise increasingly complex theories and technologies; and we evolve an understanding of natural principles; this, which all parties accept, is an informative, psychological sense of 'evolution'. The *fourth* sense of evolution, at least as near to the original Latin as the other three, is the spiritual usage; immaterial spiritual evolution, unacceptable to materialists and unknown to physical science, is at the very heart of holism; in this voluntary sense of evolution practitioners cast off material attachment, evolve and merge into the *Logos*; evolution (or, perhaps better, centripetal involution) of the soul is their great business; their aspiration is to unite with The Heart of Nature.

evolution pre-Darwinian: minority/ anti-mainstream pre-Socratic snippets and sense-based Epicureanism lionized by interpretations of post-18th century materialists; virtually undetectable eccentricity in Chinese, Indian and Islamic literature; natural selection treated by creationists al-Jahiz and Edward Blyth; Buffon, a non-evolutionist, addressed 'evolutionary problems'; Lamarck (evolution by inheritance of acquired characteristics); hints in poem by Erasmus Darwin.

evolution Darwinian: mechanism - natural selection; major tenets - common descent (inheritance), homology and 'tree of life'.

evolution: neo-Darwinian/ synthetic: as Darwinian, except synthetic theory adds random mutation as the mechanism for innovation; also adds a mathematical treatment of population genetics and various elements (e.g. geno-centric perspective) derived from molecular biology.

evolution: post-synthetic phase: natural selection and random mutation are acknowledged as mechanisms insufficient to source bio-information; post-Darwinian evolution invokes mechanisms from hypotheses such as *NGE* (natural genetic engineering) and 'evo-devo'; holistic possibilities also address the origin of complex, specified and functional bio-information.

existence: which 'stands out' from background 'nothingness'; the apparently divided universe; seemingly disparate, finite things; all motion/ change/ relativity; all psychological and physical forms and events.

F

field: any extent wherein action either physical or metaphysical but of a certain kind occurs e.g. field of battle, influence of mind or magnetism; the scientific definition is limited to a collection of numbers varying from point to point - such as a scalar field of contours on a map - or numbers with direction - such as a vector field showing speeds and directions of wind.

first cause(s):   first cause is first motion in a previously undisturbed, pre-conditional field; normally considered the first impetus to a chain of events; such 'horizontal causation' is complemented by the 'vertical causation' of Natural Dialectic's hierarchical view of cosmos as explained in Chapter 2.

First Cause Psychological is Archetype, Potential Informant or (see Chapter 5: Top Teleology) *Logos*; attributes of this Primary Source and Sustenance of Creation include omnipresence, omnipotence and omniscience.

first cause physical is called potential matter or archetypal memory; as the secondary source of creation it precedes physical phenomena; as such it is, transcending physical appearances, metaphysical; this 'physical nothingness' is therefore, paradoxically, the source of everything composing astronomical cosmos; it consists of their internal being as opposed to external manifestation or their essence as opposed to quantum or bulk state appearances; its void, with respect to the presence of finite phenomena, appears infinite; 'holographic' attributes of immanent archetype, the primary informant of our non-conscious, energetic universe, include omnipresence and omnipotence; also check the Glossary for archetype and transcendent projection.

free will: free will, reflecting cosmos, occurs in three stages; first, in non-conscious, automatic matter, that is, the physical universe, there is none; second, in conscious mind freedom of will is relative; the degree of this relativity is a function of the type of an impression, that is, of the higher or lower quality of a memory, purpose, form of thought or feeling on mind's spectrum; third, the Essential Nature of Unconstrained Free Will is, paradoxically, 'constrained' by Transcendent Love.

G

gamete: sex cell with half of full genetic complement i.e. a single set of chromosomes.

gene: generally means a basic unit of material inheritance; section of chromosome coding for a protein; digital file; a reading frame that includes exons and introns; the old one gene-one protein hypothesis is incorrect; in fact, by gene splicing, a particular piece of *DNA* may be used to create multiple proteins.

genome: total genetic information found in a cell: think of the genome as an instruction manual for the construction and physical operation of a given organism.

genotype: the genetic constitution of an organism, often referring to a specific pair of alleles; the prior information, potential, plan or cause of an effect called phenotype.

gravity: in physics an attractive mass-to-mass force or warping of space-time; the influence of mass; in Natural Dialectic the term is redefined more broadly - the agency of its (*tam* ↓) downward vector includes all psychological and physical factors of materialisation; such 'gravitational' factors and their properties are listed in the left-hand column of Secondary, Existential Dialectic; they include pain, pressure, confinement, strong nuclear force, mass (sometimes called gravitational charge), electromagnetic binding, inertia, entropy, 'standard' gravity and so on; gravity might be summarised as 'negative power' or 'the principle of death'.

GTE general theory of evolution; *see* macroevolution.

H

holism: embraces material as well as immaterial (psychological) science in its compass; opposite of reductionism; the view that a whole is greater than the sum of its parts; the extra metaphysical (immaterial) ingredient is identified by Natural Dialectic as information; information implies the purposeful design, development and arrangement of contingent parts in a working system; cosmos may operate according to a Logical Norm.

hologram: a 3-d photograph made with the help of lasers; unlike a normal photographic image each part of it contains the image held by the whole.

homeostasis: vibratory or periodic control of a system to obtain balance round a pre-set norm; the mechanism of its information loop involves sensor, processor and executor; the operative cycle works by negative feedback; dynamic bio-equilibrium; psychological (nervous) and biological cybernetics; the informed basis of biological stability.

Hox **gene:** homeotic gene involved in developmental sequence and pattern; high-level co-determinant of the formation of body parts.

146

I

illusion: is the cut between illusion and delusion an illusion? illusions, apparently outside the mind, appear real; a delusion, in it, we think real; illusion is a lesser truth; set against Absolute Truth (or Reality or Knowledge) it is a relative truth; hierarchical existence is composed of relative truths that range from slightly to completely false; only in Truth, only from the perspective of Knowledge do the illusions and delusions of existence wholly disappear (see Glossary truth; also *SAS* Chapter 4 and *PGND* Chapter 16: Truth, Appearance and Reality).

information: Latin: giving form to; the immaterial, subjective element; information tells the way; it is action's precedent; nformation occurs in three distinct modes; informative *potential* is action's precedent; this potential is both the source and substrate of psychological and physical activity; *active* information is the inhabitant of its own centre, mind, whose substrate is consciousness; such information knows, feels, purposes and codifies; it recognises meaning; on the other hand *passive* information reflects active; it is stored as subconscious memory; or is fixed in the expressions of non-conscious matter according, universally, to the archetypal behaviours of natural bodies or, locally, to particular constructions by life-forms.

informative entropy: loss of information due to degradation of its carrying medium; such a medium may be metaphysical (mind) or passive and physical (for example, computer files or genetic code); and its entropy may be metaphysical (loss of memory, focus or consciousness) or physical (for example, genetic mutation); the informative correlate of such degeneration is diminished organisational capacity, meaning or thrust of original purpose.

informative negentropy: gain of informative clarity; increasingly focused, purposive specificity; associated with knowledge, wisdom, grasp of principle and pristine construction.

inversion: turning upside-down or inside-out; reversing an order, position or relationship; in a hierarchical sense inversion is allied with the reflective asymmetry of opposite poles; information outwardly expressed; pole-to-pole reversal integral to (reading) dialectical structure; various kinds of inversion - psychological, physical, chemical and biological - are referred to in the book.

K

karma: action; law of cause and effect, that is, balance between action and reaction; *equilibration* such as underlies all mathematical *equation*; a deed or event with implications of the reactions or 'payback' it provokes; fruit or result of previous thoughts, words and deeds; *karma* delivers fate and drives destiny; inexorable law of balance covering all metaphysical (psychological) and, as in Newton's Third Law of Motion, physical forms of change.

L

levity: agency of the (*raj* ↑) upward vector; dialectical converse of gravity; psychological and physical 'levitatory' forces lift or stimulate; they are listed in the right-hand column of Secondary, Existential

Dialectic and include light, heat, excitement, dematerialisation, release, negentropy, focus of interest, affection and so on; physically, levity includes anti-gravity or the intrinsic property of matter's absence, space; in physics, the opposite of gravity, that is, the buoyant influence of masslessness (i.e. space) now termed 'dark energy'; generally summarised as 'positive power' or 'an unbinding principle of lveliness'.

logic: analysis of a chain of reasoning; principles used in circuitry design and computer programming; 'normative reason' relates to the basic axiom(s) of a given standard e.g. *bottom-up* materialism or *top-down* holism; three main logical thrusts are: (1) *inductive* (premises/ observations supply evidence for a probable/ plausible conclusion) as in the case of experimental science working *bottom-up* from specific instances to general principle: (2) *abductive* (best inference concerning an historical event): and (3) *deductive* (conclusion in specific cases reached *top-down* from general principle): two pillars of logic are holism and materialism; holism employs mainly deductive/ abductive operations and a Logical Norm; materialism tends to inductive/ abductive operations whose axis is non-conscious force and chance.

***Logos*:** First Cause; The Informant; Prime Mover; Initialising Code; Causal Motion in the form of vibration that sustains creation's conscio-material gradient; along this gradient *Logos* shows aspects of vibration we call, in physical terms, sound and light. *Logos,* transcending mind, is Conscious; therefore, _Who_ is *Logos*?

M

machine: def: a body or assemblage of bodies used to transmit or modify force and motion to produce an intended result; oblivious, physical entities *per se* lack intention and therefore cannot compose a machine (except in the Newtonian idea of the physical universe of astronomy and cosmology made by a Creator); machines conform to physical constraints but always require informative as well as energetic input/ output; the (prior) informative input involves will-power, reason, logic and often blueprint and code; such instruction *always* demands the coherent motion of interlocking parts towards a specific outcome; on the other hand, the behaviour of physical objects and forces is reflex, goal-less and lacking reason; in short, since cause of a machine is always purposeful, it is a teleological construction; such teleology applies to all physical bodies, including codified, biological ones (such as your own).

macrocosm: the physical universe of astronomy and cosmology; dialectically, the whole of existence (i.e. both universal mind and universal body) as opposed to individual, microcosmic objects and events - including the human body.

macro-evolution: large-scale, non-trivial, transformative evolution; unlimited plasticity; process of common or phylogenetic descent alleged to occur between biological orders, classes, phyla and domains; includes the origin of body plans, coordinated systems, organs, tissues and cell types; unexplained by mutation, saltation, orthogenesis or any known biological mechanism; sometimes called '**general theory of evolution**' (*GTE*); macro-evolution, an extrapolation from Darwinian micro-evolution vital to sustain a 'progressive' materialistic mind-set, is conjecture.

148

manas: pre-scientific term; conscio-material band/ grade - both conscious and subconscious bands of mind; 'mind-stuff'; 'clay' moulded by the the hand of thought and perception; metaphysical 'material' on which direct formative action of *chitta* occurs; 'film' on which the perceptions of mind are developed and, at the same time, 'screen' on which they are perceived; receptor for sense impressions and storage silo of such impressions as 'seeds' or 'files' of subconcious memory; substance of archetypal field, in other words, of universal archetypes (cf. typical mnemone); mental form and energy.

mantra: archetypal symbol; psychological transformer; authorised form of words repeated to exclude other thoughts; examples include the 'Hail Mary', '*Om Mane Padme Hum*' and, materialistically, 'evolution made...', 'in time nature designed...' or similar incantation.

maya: partial truth; world of forms and forces; illusion that changeful cosmos is the ultimate reality; motions and perceptions composing *maya* are thus, set against Absolute Truth, more or less unreal; becoming wise to the nature of *maya* yields liberation from its cosmic veil; *maya* is also known as 'not-truth' i.e. creation as opposed to Creator; its complete relativity includes, of course, one's own body; it also includes one's mind so, in this case, how can Truth ever be known? Or is everything at most a partial truth? Those who, as materialists, believe that non-conscious energy is the ultimate truth behind all things believe that the Fact of a Conscious Creator is itself a delusion.

matter-in-practice: bulk, bonded matter including all molecular-based substances; gross matter; external appearance.

matter-in-principle: quantum phase; particles and forces; subtle matter; internal cosmic drivers.

meditation: Lat. *medius*, middle; coming to Centre; mind and body dropping away.

meiosis: shuffling the information pack: variation-on-theme; mechanism for the production of haploid gametes; genetic postal system for sexual reproduction.

metabolism: body chemistry.

metaphysic: = non-physical/ immaterial/ psychological/ unnaturalistic (if physic is equated with natural); everything non-physical; physically expressed as specific/ intended arrangement/ behaviour of materials; physical behaviour reflects a subtle, non-physical blueprint; involves element of information; also involves symbol/ code/ abstraction/ logic/ reason/ mathematics; and message/ meaning/ goal/ teleology; and consciousness/ mind/ life/ experience/ feeling; and also morality/ force psychological/ emotion; involves innovation/ creativity/ art/ invention/ aesthetics.

microcosm: an entity that reflects the universe by containing all its basic constituents. Used especially of the human state where it may refer to both mind and body or, in a purely physical context, body alone.

micro-evolution: misnomer; non-progressive, small-scale variation within a species or, more broadly, between strains, races, species and genera; limited plasticity; *variation/ adaptation within type*; trivial Darwinian changes that may occur by natural selection/ ecological factors acting on genetic recombination, adaptive potential or mutation; also called '**special theory of evolution**' (*STE*), *STE*/ variation is a fact.

mitochondrion: organelle in eukaryotic cells containing the apparatus for aerobic respiration.

mitosis: conservative copying and delivery of genomes in cell division; genetic reprinting; genetic postal system for asexual reproduction.

mnemone: a division of memory whether individual or universal: an individual's two divisions are *personal mnemone* (likened to a working cache or data store) and *typical mnemone* (likened to a *ROM* or an operating system); typical mnemone is, in effect, a program consisting of three subroutines - *signal translation*, *instinct* and *morphogene* (for more information see *SAS* Chapters 15 - 17); it is also a synonym for natural, archetypal memory in universal mind; in short, it is a body's *metaphysical DNA.* Further than the character of each bio-type of organism it also includes the 'instinct' of matter (i.e. cosmos). Natural Dialectic's definition involves no 'cultural' connotation whatsoever and is thus wholly distinct from evolutionary psychology's use of the word.

morphogene: one of three sub-routines of typical mnemone or archetypal memory relating to physical construction; morphological attractor; the component of subconscious mind associated with electrochemical function and thereby body; just as you might not guess from the picture on your TV screen or object from a 3-d printer the nature of the electromagnetic messaging that creates it so you might not guess a body's shape from its *DNA* or the messaging agent that links archetypal mind with body; morphogene is the dominant aspect of mind in unconscious organisms such as plants or fungi.

morphogenesis: the development of biological structure; more generally, the production of physical form (of which morphology is the study).

mutation: accidental change to genetic code.

mysticism: quite different from objective, it is the subjective science; not philosophy, religion or opinion but practice to achieve communion with natural, inner, immaterial truth; esoteric as opposed to exoteric, materialistic discipline; 'science of the soul'; as gyms and physical action are to athletes so meditative exercise and psychological stillness are to mystics; involves psychological techniques to achieve a clear, rational goal - purity of consciousness and thereby understanding of the fundamental nature of the informative principle, mind; since life is lived in mind a mystic seeks consummate knowledge of life's source and sanctum, that is, communion with its deathless heart; adepts were, are and will be 'Olympian' meditative concentrators.

N

natural law: the automatic, reflex and mathematically describable behaviour of a physical entity; likewise the repetitive nature of its interactions with other entities.

natural dialectic: dialectic is a method of discourse between two (or more) people holding different points of view about a subject, who wish to establish the truth of the matter guided by reasoned arguments; such dialogue has been central to European, Buddhist and, in the Taoist treatment of opposites, Chinese philosophy since antiquity. In Europe it was made popular by Plato and Aristotle; also William of Ockham, Thomas Aquinas and latterly, with differing shades of usage, Hegel, Kant, Marx and others. At medieval Oxbridge it was taught under the heading of logic along with rhetoric and grammar; professors examined this threefold 'trivium' in dialectical discourse with their student sat on a three-footed stool (or tripos); if successful he was awarded a degree. This book's Philosophy of Natural Dialectic (effectively a Theory of Complementary Opposites) uses its own robust, binary vehicle with which to make comparisons, connections and explanations; its polarities represent not only human but the cosmic spine; or, if you like, the system generates a muscular body of philosophy that accurately reflects the order of the cosmos. It generates an abstract, metaphysical machine, tight-knit, well riveted by bolt and counter-bolt, the simplest working model of the universe; see also unification (see also Dialectic).

naturalistic methodology: also known as 'methodological naturalism', this strategy is, strictly, not concerned with claims of what exists or might exist, simply with experimental methods of discovering physically measurable behaviours; thus only materialistic answers to any question (e.g. how biological forms arose or the nature of mind) are deemed 'scientific' or 'scientifically respectable'.

negentropy: opposite of entropy; lowering of entropy; expression of the (*raj*) upward-pointing cosmic fundamental closely coupled with stimulus, dissolution and dematerialisation; a measure of input, cooperation or synthesis; motive/ fluidising aspect of an energetic, material or conscious gradient; gain of energy, configurative order, information or consciousness in a system; when used in terms of information negentropy involves gain in order or understanding of principle from which different actualities derive; a measure of the amount of concentrated/ conceptual information, specific, intentional complexity or conscious arrangement in a system; a natural and essential property of mind.

nirvana: state of enlightenment; 'non-condition'; nirvana is devoid of existential motion; extinction of existence (i.e. perpetual change) leaving Essence Alone; pure soul; psychological super-state; Buddhists call such transcendence non-self or the Formless Self.

non-existence: where creation = formful existence, non-existence is formless; the polar opposite of physical space and time is Transcendent Potential; such pre- or super-existential formlessness is non-existent; Absolute Non-Existence is Essential; however relative non-existences of two kinds also occur; the first kind is metaphysical/ subjective and therefore psychological; it involves the absence of a specific psychological form or event; unconscious oblivion is one such non-existence; the second kind involves the local absence of a possible physical event (an object is a 'slow event'); impossibilities are non-existences but imaginations of non-existence (including symbolic abstractions, hypothetical entities, physical absences, absolute emptiness and the number zero) exist; furthermore, the nothingness of space and time, the zero-point of calculus and zero's empty set together constitute the basis of physical science and mathematics.

noumenon: implicit, metaphysical cause of an explicit, physical phenomenon; from the Greek word *nous* meaning mind; see also archetype.

nucleic acid: *see DNA and RNA*

nucleotide: basic, triplex unit of nucleic acid polymer; monomer composed of phosphate and sugar (the 'paper' part) and base (the 'ink letter'); letters' of the genetic alphabet are (G) guanine, (C) cytosine, (A) adenine and (T) thymine. In *RNA* thymine is replaced by (U) uracil.

nucleus: centre, heart, creative core; informative *sine qua non*; psychological nucleus is consciousness or (in formful aspect) mind; atomic nucleus, made of protons and neutrons, is a centre of mass determining electron configuration; biological cell nucleus is the instruction centre of a cell containing *DNA* and nuclear operating machinery; nuclear is critical.

O

object: a slow, although energetic, event; apparent fixation.

Om: universal sound, fundamental reverberation, basic creative agent; it is the creative/ informant energy of vibration; and it propagates through the field of consciousness *via* mind to anti-Source or field periphery/ sink viz. the created/ passive/ reflex energy of matter; it is sometimes spelt *Aum*, a Sanskrit word whose Semitic transliterations are Am'n, Amin and Amen; see also First Cause, *Logos*, *Kalam*, *Shabda* etc.

order: regular, regulated or systematic arrangement; organisation according to the direction of physical law; passive information by which things are arranged naturally (with predictable but non-purposive complexity) or purposely (with innovative or specified complexity); mind, generating specified complexity in the order of its technologies and codes, actively informs; the orders of mind are meaningful, the orders of matter lack intent; see also cosmos.

P

PAM, PAND, PCM **and** *PCND*: philosophical gambits; see Primary Axioms and Corollaries.

phenotype: the effect of causal potential; result of the development of prior, informative 'egg'; outward expression of inner plan; sensible appearance of an organism as opposed to its genotypic scheme: the whole set of outward appearances of a cell, tissue, organ and organism are sometimes called a phenome (*cf.* genotype/ genome).

photosynthesis: process by which inorganic carbon is introduced to the biological zone and energetic sunlight fixed as a crystalline molecule of storage, a sugar called glucose.

phylogeny: evolutionary history; relationship of shape (morphology) and similar structure (homology) are construed to demonstrate common, evolutionary or phylogenetic descent.

potential: poise; latent possibility; contextual capacity to develop; potent non-action that precedes any particular action or creation; in science potential energy is defined as the energy particles in a system (or field) possess by virtue of position/ arrangement; gravitational, electrical, electro-chemical, thermo-dynamical and other kinds of potential are recognised; in dialectical terms mind precedes matter, information precedes the pattern of material behaviour; information is energy's pre-requisite potential; in this case *informative potential* involves two conditions; firstly, a pre-existential/ essential state of psychological potential; secondly, a pre-material, metaphysical fact of potential matter, archetype or laws of nature; if potential's pre-active equilibrium is related to the voltage of a full battery then aspects of psychological 'voltage', whose currents drive intentional behaviour, are purpose, will and plan; *see* Chapter 2 diagrams re. potential and first causes.

potential matter: see archetype.

prakriti: is a Sanskrit word. What does it mean? Before creation is a pre-creative potential. In this non-existent but essential unity there is only complete, motionless balance but with the commotion of creation springs vibration (see Glossary: *Om* and *Logos*). In a binary creations two currents, radiant and anti-radiant (constrictive) interact. As regards cosmic fundamentals we name one (↑) the positive, dematerialising or liberating power and the other (↓) negative. The latter's current of materialisation is the objective energy at the root of form. It is anti-ascent (descent); its influence tends to reduce, constrain, press down and fix both psychological and physical forms. Lucifer turned his back on God, *Brahm* polices cosmic law and, casting his universal net, the devil entangles souls from their return to Source. In Hinduism that which creates a sense of separateness or form is called *Maya*. Whatever you call it, this negative influence keeps creation turning. Thus, complementing the positive it is critical for creation. It effectively 'programs' and gives form to ideas of the positive power. What expression lacks shape? In other words, *prakriti* may be likened to the material component of the binary 'conscio-material gradient' of creation. It is also called primordial matter, ylem, 'nature' or 'the full range of nature' and can be likened to a contractive spectrum - an anti-spectrum of radiance or the materialising power of darkness. Non-conscious *prakiti* is the material component of the (↓↑) conscio-material gradient and is concentrated (or fully expressed) in the forms of energy that science studies. Regarding this lowest band it has been dubbed 'the mother-field' of our physical universe.

prana: Energy vibrates with two main aspects - sound in a medium and light (also through a vacuum). In each case physical science employs the same unit of oscillation or cycles per second (Hertz); and, while sound is scaled, electromagnetism is measured by spectra. These spectra are divided into arbitrary groups of frequencies called wavebands. While the tiny waveband upon which earthly life depends life is called the visible light band, others include ultra-violet, infra-red and radio frequencies. So with *prakriti* whose subconscious, *psychosomatic waveband is called qi, ch'i or prana. Prana* in yourself is associated, as in the yogic practice of *pranayama*, with breath and thereby life of material body; and also identified, by some, as 'metaphysical electricity', psychosomatic energy or, by others, as material electromagnetic light and oxygen (specifically, negative ionic charge). *Prana,* often dubbed life-force, is the archetypal energy of subconsciousness; its 'vibrant electricity' powers the operations of subconscious mnemones, that is, all *PSI* (psychosomatic) traffic between

153

conscious and subconscious grades of mind along with non-conscious objects such as atoms or stars. As such it is like a 'VLF' or 'radio' aspect of the Logical Sustenance, Essential First Cause or Word that undergirds all motion, that is, all creation. We call this 'infra-mental' radio band *'potential matter'*. In other words, as life depends on sunlight, so perpetual atomic motion is supported by its never-ending underlying vibration. As the band of visible light is divided into a rainbow of colours, so five prismatic sub-divisions of *prana* work through *chakras'* (*see* Glossary). The distributors of psychic energy correlate with the expression of a range of elemental characters in bio-physical systems such as digestion, water control and distribution of oxygen. These are 'tuned' to different degrees, including a point of absence, in the range of terrestrial life-forms. In humans (*see* Chapter 3: Informed Man) all are active starting at the top from *'ajna chakra'* or third eye behind the forehead; this, in turn, is subservient to *'sahasrara'*, the 'thousand-petalled lotus' supplying 'voltage' (and thereby current) to the 'grid' that sustains cosmos. Physicaly and biologically *pranic* frequencies resonate with a grid of aforementioned nodes, well-known to yoga and arranged down the spine, each broadcasting a frequency appropriate to its body area using a network of *nadis* or meridians identified by medical acupuncture; being metaphysical the *pranic* system cannot be physically tested by empirical, scientific experiment but only by inference (e.g. a cure); for this reason some proponents of occidental medical science dismiss *pranic* mechanisms of the mind as 'pseudoscientific' and, having thus 'rationally' condemned, proceed to narrowly dismiss the broader immaterial fraction of holistic order wherein such components play a crucial part; each of the five 'coloured' bands is associated with a *chakra* and, for health, they need to work in unobstructed, mutual balance; briefly described, *prana* (↑) is an upward, stimulatory or vital force; at the third eye (*ajna chakra*) its bio-function is associated with sensation/ discrimination between individual objects, events and qualities; at the heart centre, with vital respiration; and the throat centre's *udana* coordinates nervous activity concerning balanced mobility and buoyancy; it is also associated with 'raising of a so-called *kundalini* force' normally dormant at the base of spine; *samana*, whose centre is the solar plexus, treats digestion and heat production (metabolic respiration); the function of sacral *vyana* is resupply, maintenance and preservation; finally, gravitational *apana*, the downward (↓), materialising antithesis of *prana*, is involved with the regulation of waste and expulsion of bodies to earth; in living organisms (including you and me) wireless *pranas* are, as a whole, identified as a metaphysical spectrum of life-forces animating the typical mnemone and through this psychosomatic linkage a physical body.

Primary Axiom of Materialism (*PAM*):   all objects and events, including an origin of the universe and the nature of mind, are material alone; cosmos issued out of nothing; life's an inconsequent coincidence, a fluky flicker in a lifeless, dark eternity.

Primary Axiom of Natural Dialectic (*PAND*):   there exists a natural, universal, immaterial element - information; immaterial informs material behaviour; a conscio-material dipole that issues from First Cause informs

154

and substantiates both mental (metaphysical) and physical creations; there is eternal brilliance whose shadow-show is called creation.

Primary Corollary of Materialism (*PCM*):   the neo-Darwinian theory of evolution, that is, life forms are the product, by common descent, of a random generator (mutation) acted on by a filter called natural selection; such evolution is an absolutely mindless, purposeless process; the *PCM* is a fundamental *mantra* of materialism.

Primary Corollary of Natural Dialectic (*PCND*):   the   origin   of irreducible, biological complexity is not an accumulation of 'lucky' accidents constrained by natural law and death; forms of life are conceptual; they are, like any creation of mind, the product of purpose.

prokaryote: non-eukaryote;   bacterial   type   with   little   or   no compartmentalisation of cell functionaries.

promissory materialism: belief sustained by faith that scientific discoveries will in the future justify/ vindicate exclusive materialism and, as a consequence, atheism; may involve a call to progress towards the technological provision of its 'promised land'.

protein: factor made from a specific sequence of amino acids to perform a specific task; 'informative' protein includes some hormones; skin, hair, bone, muscle and other tissues are made of 'structural' protein; 'functional protein' called enzymes mediates all stages in cell metabolism, that is, it catalyses all biochemistry; 'informative'-protein signals/ triggers/ communicates cell activities.

***PSI* (psychosomatic interface):**   psychosomatic border; the level of mind-matter interaction; bridge between metaphysical and physical dimensions; potential matter; 'gap of Leibniz'; 'fit' of mind to matter; point of linkage between subconscious mind and non-conscious matter; gearing between instinct/ archetype and the behaviour of material objects and energies; as in the case of physical law, psychosomatic influence is both general in potential and local/ specific in engagement.

psychological entropy: a measure of loss of concentration, focus of attention or consciousness; loss of 'mental energy' or aptitude; the drop from waking to sleep; loss of knowledge, information or sensitivity; the gradient from intelligence through stupidity to oblivion; an expression of the (*tam*) downward cosmic fundamental in mind; a tendency predominant in lower, egotistical or selfish mind; increasing level of ignorance, anguish or immorality; loss of integrity, psychological disharmony or disintegration; see also *information entropy*.

psychological negentropy: a measure of gain in order; an increase in concentration, focus of attention or consciousness; gain in sense of purpose, 'mental energy' or aptitude; the rise from sleep to waking, 'dark to light' or unhappiness to happiness; gain in knowledge, information or sensitivity; the gradient of learning and spiritual evolution; an expression of the (*raj*) upward cosmic fundamental in mind; a tendency predominant in higher mind; increasing level of contentment, understanding and the natural morality of happiness; the ascent towards psychological radiance, harmony and integration. The converse of psychological negentropy involves *entropy of information*.

155

psychosomasis: operation across the psychosomatic border; mind/ body interaction; the one-way, morphogenic imposition of archetypal pattern on *physicalia*; the two-way exchange of information in sentient organisms through the agency/ medium of subconscious patterns.

Purusha: pre-scientific term; conscio-material band/ grade - conscious; Pure Consciousness, Source of Life, Subjectivity and Creativity; Universal (*Sat*) Potential; boundlessly pre-active,and in creative action Prime Mover, First Cause or Top Governor on the universal scale and order of creation; given many other names in many languages; ultimate subject of worship, praise and love.

Q

quantum: minimum discrete amount of some physical property such as energy, space or time that a system can possess; quantum theory states that energy exists in tiny, discontinuous packets each of which is called a quantum; an elementary discontinuity; an elementary particle e.g. photon or electron.

quantum level: matter-in-principle; 'internal', 'causal' or 'subtle' matter; the vibrant or energetic phase of physical organisation; zone of sub-atomic particles and forces; step (on cosmic ziggurat) between potential and bulk matter whose aspect is sometimes extended to include atomic and molecular interactions; small-scale substance underlying large-scale, sensible appearances.

R

raj: (↑) upward, levitatory or stimulatory cosmic vector.

reductionism: opposite of holism; the materialistic view that an article can always be analysed, split up or 'reduced' to more fundamental parts; these parts can then be added back to reconstruct the whole; a whole is no more than the sum of its parts.

religion: except in the case of materialistic species, an attempt, encrusted with metaphor and ritual, to describe, explain and interact with Essence; this Essence, labelled many ways, is the Indescribable Heart, the Transcendent Nucleus of existence; etymology of the word is debated between Latin *relegere* (review) and *religare* (bind); this latter includes the sense of *yoga* which also means joining, yoking or connection; *religio* means dutiful and meticulous observance; currently religion means world-view, mind-set or basic faith; whether of materialistic or holistic belief, it involves the non-negotiable substance of an individual or community's truth - notably as regards origins; antagonism between holistic practice and the naturalistic methodology of science is, because the couple deal with separate but complementary physical and metaphysical dimensions, flawed; a materialist/ atheist 'binds meticulously' to an evolutionary mind-set, a holist to a Primary Source whose influence is, stepped through the levels of creation, everywhere; in the case that self-deception is crucial to successfully deceiving others which, holism or materialism, is the religion that is ultimately true?

resonance: the tendency of a body or system to oscillate with a larger amplitude when subjected to disturbance by the same frequencies as its own natural ones; thus a resonator is a device that naturally oscillates at such (resonant) frequencies with greater amplitude than at others;

resonance phenomena occur with all kinds of vibration, oscillation or wave; their sorts include mechanical, harmonic (acoustic), electrical (as with antennae), atomic and molecular.

respiration: the controlled release of energy from food.

RNA: a single-stranded nucleic acid polymer employed in three different forms (*m- t- and r-*) during the process of protein synthesis; in computer terms might be likened to a portable memory stick as opposed to *DNA*'s hard drive.

S

sanskara: character trait; groove, habit, obsession or repetitious mode of thought proportional in depth to the intensity of desire, force of impact or impression that created or sustains it.

samsara: existence, phenomena, the place of cycles and, therefore, reincarnation; non-essence or, in Buddhism, what is not *Nirvana*.

sat: 'top', key or essential cosmic fundamental; 'vector' of balance, neutrality.

science: Latin *scire* (know); knowledge; commonly understood as the practical and mathematical study of material phenomena whose purpose is to produce useful models of the physical world's reality; in 19th century the word replaced 'natural philosophy'.

scientism: a philosophical face of official, *de facto* commitment to materialism; today's majority consensus of what the creed of science is; an -ism born of *PAM*; a faith that all processes must be ultimately explicable in terms of physical processes alone; like communism, a one-party state of mind; a doctrine that physical science with its scientific method is ultimately the sole authority and arbiter of truth; a set of concepts designed to produce exclusively material explanations for every aspect of existence, that is, to colonise each academic discipline and build its intellectual empire everywhere; 'scientific fundamentalism' closely allied, when expressed in social and political terms, with 'secular fundamentalism', sociological interpretation of behaviour and the fostering of a humanistic curriculum.

secular fundamentalism: *PAM* as applied to the worlds of nature and of human society.

secularism: concern with worldly business; lack of involvement in religion or faith; secularism is generally identified, as defined by the dictionary, with materialism; for a secularist the ultimate arbiter of truth is human reason - ideas are open to negotiation so that even morality is relative; however many liberal agnostics, atheists and humanists argue that their metaphysical, philosophical system also embraces so-called 'universal' moral values and, as opposed to zealotry or the logic of evolutionary faith, a liberal politic of 'philosophical live-and-let-live'.

siddhi: marvellous, miraculous or 'super-natural' psychic ability that, at the point a practitioner actually masters it, becomes natural.

STE: special theory of evolution; *see* microevolution.

stereo-computation: stereochemistry involves study of the relative spatial arrangement of atoms in molecules; in biology a 1-D line of informative code (whose 3-D constituents bear no figurative relationship with their informed product) give rise to relative 3-D spatial arrangements at all levels from molecular to systemic and whole-body; such targeted generation may be termed bio-logical stereocomputation.

sub-state: *opp.* super-state; impotence, discharge, exhaustion, final stage in the expression of potential; fixity; non-conscious base-state; state 'below/ subtendence; extreme negativity/ (*tam*) condition.

super-state: potential; source of possibility; causal metaphysic/ archetype; state 'before' or 'above' subsequent expression; immanence; transcendence; precondition; (*sat*) priority.

T

tam: (↓) downward, gravitational or inertialising cosmic vector.

tanmatra: pre-scientific term; possible conscio-material band/ grade - subconscious band of mind; 'simple elements' or devices that facilitate the transfer of sensory information (sound, vision, taste, touch and smell) from organs and their nervous electical coding; as such *tanmatras* are seen holistically as components of the typical mnemone and thus involve the least metaphysical, most nearly physical bands of mind; as with *chakras* and *prana*, their operations regarding pschosomasis (*see* Glossary) involve what Ayurveda calls *koshas. Koshas,* which are sheaths or bodies representing five 'layers of being' (as chapter 3: Wireless Man illustrates). Your body is, like the universe itself, composed of each of these; the subconscious sensory *tanmatras* are located in the *Pranamaya* sheath (*H. archetypalis*) whose base touches the physical *Annamaya sheath (H. electromagneticus)* most subtly expressed as charge in electromagnetic (or, in sound aspect, sonic) networks. Ayurvedic ideas (not least as regards the 'five elements) were known to and certainly influence Classical philosophy (although current scholarship, perhaps ignorant of Ayurveda, fails to make the link). Are *tanmatras*, traditionally thought of as mental ideas, the same as Platonic ideals? Or, so crude is the apprehension, are the latter simply abstract forms of geometry? Put a slightly different way, *tanmatric* apparatus represents a stage in the hierarchical, orderly two-way traffic of incoming (matter to mind) or outgoing signals (mind to matter) across either an individual or universal psychosomatic border; this apparatus, called typical mnemone, regulates image, quality and form and thus represents a qualitative aspect of matter and is the 'device' that allows mental grasp of physical sensations; as such, *tanmatras* are an instrument of potential matter, the archetypal processors of image; however, as structure to function, as mechanism to its fuel of *prana, tanmatras* are forms of 'bridge' between the so-called separation between physical and metaphysical phases of creation; this 'zone' is where numinous cloud 'collapses' into the 'drops' of individual physical phenomena (a 'phenomenon' means an 'appearance'); such 'condensation' represents the transit-point where mind and matter meet and, as such, *tanmatras* are seen as *components of secondary first cause or physical archetype*; or, again, they function as filters for orderly focus of a quantum factor, electronic charge which, in turn, under the influence of gravitational and perhaps dark and yet other unknown forces, become the gross atoms and aggregates of bulk matter that are studied by physics and chemistry. As such they are, in conjunction with 'metaphyical electric' (*prana*) as their power source, the pre-physical mechanism that expresses, in the physical vacuity of archetypal field, the quantum fundamentals of material phenomena. As an element of wireless mind, *tanmatras* transmit 'sounds' (vibrations or frequencies), whether simple, single or in 'complex opera' that codes for bio-symphony. They are vibrant metaphysical realities that translate into cymatic messages (check Index: Chladni); and resonant frequencies that excite the emergence and maintenance of physical forms and behaviours; in short, these subconscious instruments are the final agents in the initial creation of physical shape and force; cooperant *tanmatras* and *pranas* substantiate the forms of energy whose materialisation is known to quantum physics; and the bonded or aggregate fixation of quanta yields the chemistry, physics and biology of condensed/ bulk matter; inside out, inner information supports outward operations; from subtle to gross, *tanmatras* iteratively, correctly support the way our starry cosmos works.

tattwa: pre-scientific term; possible conscio-material band/ grade - both subconscious and physical bands; seems to mean 'that-ness' or 'not-self'; by nature, using their intellect/ *buddhi*, philosophers seek to analyse, categorise and argue so that, in the case of *tattwa*, the number of items listed by them in their teachings varies considerably according to tradition; basically, however, it amounts to a 'catch-all' description of human, animal and inanimate condition; five well-known *tattwas* correspond to Samkhya, Greek, Latin and medieval European elements of ether, fire, air, water and earth; on the subconscious side of *PSI* (see Glossary), in archetypal memory of universal mind, these correspond to five *tanmatras*, five *pranic* 'notes' and five lower *chakras*; and on the physical side to five informative senses and five energetic organs of action called *indriyas*; the five elements/ states of matter are variously defined; ether (nothing whatsoever to do with material objects or events) is space, upper sky: also archetypal potential (which, being metaphysical, is physically unseen); subconscious etheric frequency is related to the throat chroat *chakra*, seat of dormancy; there follow gas (air *tattwa* related to breath, oxygen and heart *chakra*), energy (fire *tattwa* giving stimulus for change, heat and light whose hub is the solar plexus), liquid (water and its osmoregulatory and waste expulsion systems) and solid (the earth *tattwa* of related to bio-mass and its sense of pressure/ touch whose *chakra* rests at the supportive base of the spine).

teleology: the doctrine that there is evidence of purpose in nature; doctrine of non-randomness in natural architecture; doctrine of reason ('for the sake of', 'in order to', 'so that' etc.) and intent behind biological and universal design.

third eye: place where you think; point of metaphysical focus between and behind the eyebrows, that is, just above the physical eyes; HQ/ seat of mind beyond the sensory world; cosmological eye-centre; gate through which meditative concentration can pass; single way that leads within.

transcendent projection: creative projection from intrinsic Source; see Primary (Metaphysical) and Secondary (Physical) First Causes from Chaps. 3 and 4; voluntary issue from potential mind or involuntary from potential matter; in other words, transcendent metaphysical projection is through Causal Archetype and transcendent physical projection through mnemone (*see* Glossary & Index).

psychological: projection is through Causal Archetype called *Logos*, Holy Name, *Sat Nam*, *Om* and many other names. This projection is from

Alpha Point, the first and highest cause of creation; the First State of Super-consciousnessis Psychology's Essence.

physical: as Chapter 4 explains, physical first cause is projection through potential matter or base archetype in universal mind; such memory is the source of cosmo-logical language; it involves the orderly expression of energetic forces and particles; linkage, as Chapter 3 suggests, by psychosomatic resonance; emergence of physical phenomena from archetypal noumenon, that is, from unseen potential; an instantaneous 'miracle' that issues from 'within' non-conscious physicality; transcendently emergent, finely tuned expansion from 'inner' metaphysic into 'outer' material/ natural law; physical nothingness from whose prior pointlessness (or 0-dimensional singularity) all points began; cosmic seed whence, *ex nihilo*, the world developed; projection whose appearance, once physical, is visible and perhaps described but certainly not explained by big bang theory; transcendent projection of archetype is possibly, to the constrained sensory and intellectual states of human mind, ultimately incomprehensible; its invisible dynamic, the practice of materialisation, may remain a fact beyond material understanding. for references involving more detail about psychological, physical and biological projections see *SAS* or *PGND* Glossaries.

biological: if matter is developed memory (*see* chapter 5) conceptual biology expands this theme; biology is based on information in the form of code and codification, demanding forethought, is a product of mind; intrinsic archetypal program is reflected by extrinisic chemistry of a body which, in terms of Natural Dialectic, is passive information. *Based on information, the basis of biology is metaphysical.*

truth: what's correct or accurate; a universal truth substantiates, as source and sustenance, all things; man's holy grail is truth; *top-down*, within a hierarchically devolved construction (such as a machine, cosmos or other working system) truth is vested both in its source and physical appearance; in this case, viewed at their own level, psychological (subjective) or physical (objective) facts appear self-evident, obvious realities; however, they are relative or lesser truths when set against Truth Absolute; from different heights on a mountain slope the climber's perspective changes; so with Mount Universe, whose Peak represents Whole Truth; in the light of Peak Reality all existence seems real but is composed of relative illusions; it is a shadow play dependent on The Independent Source; man is born to seek and find this Essential Source, this Alpha Point, his Origin; see also Glossary: illusion.

U

unification: simplification; details are unified by their working principles, themes or programs; better to perceive intrinsic principle is to simplify or unify an understanding; progressive unification of forces is the grail of physics: Clerk Maxwell unified electricity and magnetism; electroweak or *GSW* theory brought in the weak nuclear force; now the goal is to include the strong nuclear force (*GUT*), gravity in a super-force and show that, in essence, particles and forces are interchangeable (super-symmetry and *TOE*); Natural Dialectic, also working with the maxim 'All is One', includes what sums to a hierarchical *TOP* or Theory of Potential; potential (see *SAS*: Glossary and Index for archetype and potential) is an absolute from which variant orders of relativity derive; the equivalent of *TOP* is *CUT* (see index: cosmic unification theory); **Natural Dialectic is the vehicle of CUT, whose aim is to build cosmic connections, that is, orderly linkages towards a Holy Grail of Unification**; the Great Connector, that is, Unifier is consciousness; the subjective potential for mind is consciousness and the objective potential for matter is archetypal memory; such archetypal element unites psychology with the physics of natural science; it is the informative precondition of physical and biological form.

universal mind: cosmic grade; also called the 'mind of nature' or 'natural mind'; as a biological body is a specific though complex arrangement of universal chemicals so individual mind partakes of a particular, equally minuscule fraction of the metaphysical components of universal mind; *see* also archetype.

V

vector: existential dynamic; a vector has both direction and magnitude; it illustrates direction of travel with respect to a model or a secondary stack used in Natural Dialectic; fundamental vectors (↑ and ↓) denote relative gain or loss of information or energy; and, similarly, motion towards and from the axis, peak or source of a cosmic model; in this case, magnitude is inferred to occur on a scale between any pair of opposites, for example, the relative proportions of black and white in the grey-scale between these opposites; use of the word is general, metaphorical rather than specific; opposite members of a stack may involve metaphysical factors (e.g. love/ hate, beauty/ ugliness) as well as physical; thus its spectra do not necessarily concern physical motion or mathematical calculation; its 'field of relativity' extends beyond non-conscious elements; in this respect a Dialectical vector is similar in principle but not the same as that defined by physics or biology; could not Natural Dialectic, a Theory of Complementary Opposites, also be called a Theory of Relativity (Mind as well as Matter) with Absolution? It is certainly a theory of change; but its (↑↓) vectored transformations, psychological or physical, register directions of informative as well as energetic/ material activity.

Vitruvian man: where art meets science Leonardo's 'universal man' demonstrates an architectural symmetry, excellence of composition and, microcosm unto macrocosm, a reflection of the universe; Da Vinci's connection, from his notebooks, is quite the opposite of Darwin's doodle (Chapter 5); if, with ratios and rationality, it demonstrates mathematical perfection then does not design of larger cosmos demonstrate it too? Natural Dialectic certainly concurs with Leonardo's logical submission.

Z

zero: zero (the number) is a metaphysical entity, one critical to mathematics; zero (the fact) means, for Natural Dialectic, nothing in two senses; in the *negative sense* it means an absence of perception (psychological oblivion) or absolutely nothing physical (as the nature of a theoretically perfect vacuum or, as naturalistically but illogically prescribed void to precede, say, a big bang); negative sense may also be construed as (*tam*) an extreme sub-state, end-point, sink or emptiness; for materialism 'absolute nothingness' may involve natural law and its mathematical (metaphysical) description; what, one may enquire, is the source of such 'eternal metaphysic', what is the nature zero-physical?; on the other hand, in a *positive sense* zero refers to source, pre-existent potential or (*sat*) higher cause-in-principle; for example, information (which is zero-physical) transcends/ precedes a course of action; information that passively governs the operation of cosmos derives from immaterial archetype.

ZPE: zero-point energy; quantum vacuum; vacuum energy of all fields in space; residual energy of all oscillators at 0°K; concept first developed by Albert Einstein and Otto Stern; intrinsic energy of vacuum; the ground-state minimum that any quantum mechanical system, in particular the vacuum, can have; remainder, according to the uncertainty principle, when all particles and thermal radiation have been extracted from a volume of space; residual non-thermal radiation; irreducible 'background noise'; 'quantum foam'; the potent, microscopic side of quantum vacuum (as opposed to impotent, macroscopic vacuum left by the apparent lack of anything); subliminal 'rumblings' of immaterial weak, strong and electromagnetic fields (called *ZPFs*); seething, jostling ferment of subliminal waves and particles in emptiness; a flux of unobservable 'virtual' matter and anti-matter that may or may not appear as the basis of observable forces such as electromagnetism, charge and perhaps inertial mass and gravity; a subtle facet of levity; the anti-gravity of dark energy (or the cosmological constant) has been postulated as a component of *ZPE*; suggested 'mother-field' support for electron orbits, atomic structure and thus the phenomenal universe.

zygote: fertilised egg.

Connections

SAS   Science and the Soul
A&E   Adam and Evolution
PGND A Potted Grammar of Natural Dialectic
AMA? A Mutant Ape? The Origin of Man's Descent
SPFP Science and Philosophy - A Fresh Perspective
RSP   The Reunification of Science and Philosophy
RN   *0* and *1* Research Notes
OW   One World Set *passim*

---

[1] *SAS*: Chapter 0: Anti-parallel Perspectives; *PGND* Chapter 1: Primary Assumptions; *A&E*: Chapter 2; *SPFP* Lecture 1: Materialism and Holism.

[2] *SAS* Glossary; *PGND* Chapter 1: Primary Assumptions; *SPFP* Lecture 1: Primary and Secondary Axioms; *RSP* Chapter 1.

[3] *SAS* Preface; *PGND*: Chapter 2.

[4] *SAS*: Chap. 1 Natural Dialectic's ABC; *PGND*: Chap. 3 Models; *SPFP* Lecture 1.

[5] *SAS*: Chap 1 Nat. Dialectic's ABC/ Glossary/ Index; *PGND* Chap 4; *SPFP* Lect. 1.

[6] *SAS passim*; *PGND* Chapters 13 to 15: triplex sciences; *SPFP* Lecture 1: simple links and Lectures 2-6: triplex, hierarchical science ; *RSP passim*.

[7] *SAS* Chapter 18; *SPFP* Lecture 1; paradigm shift.

[8] *SAS*: Chapter 0: Opposite Directions of Mind and Two Pillars of Faith; *PGND* Chapter 1: Two Pillars, A Dialogue of Faith.

[9] *SAS* Chapter 2: First Principles; *PGND* Chapter 10: Existence.

[10] *SAS* Chap. 3; *PGND* Chap. 7: A Hierarchical Perspective; *A&E*: Chap. 3; *SPFP* Lecture 2; *RSP* Chapter 2.

[11] *SAS* Chapters 3: Transcendence, 16 *passim*, 17 *passim* and 19: Conceptual Biology; *PGND* Chapter 13: H. archetypalis; *AMA?* Chapters 24 and 25; *A&E* Chapters 7 and 8; *SPFP* Lectures 2-5; also Glossaries and Indices in the books mentioned.

[12] *SAS* Chaps. 3 & 14: Subtendence and Transcendence; *PGND* Chap. 9; *SPFP* Lectures 2 & 3.

[13] but *see SAS* Chap.12 for atheism's crucial mytho-cosmology of parallel universes.

[14] *SAS* Chapter 2: Causation and *fig.* 6.1; *PGND* Chap 10; also *figs.* 11.1 and 13.5 horizontal (energetic) and vertical (informative) vectors of causation; *SPFP* Lecture 2

[15] *see* also Glossary and Index: cycle.

[16] *SAS fig.* 11.1; *SPFP* Lectures 2 and 3.

[17] *see* Glossary; also *SAS* Chapter 5 and *PGND* Chapters 11 and 12; *SPFP* Lecture 2; *RSP* Chapter 2.

[18] *see* also *SPFP* Lecture 2 for Neumann's 'physical information'.

[19] A *program* or *device* that generates randomness (e.g. card shuffle, throw of dice, random number generator or biological meiosis) is intrinsically purposive, the purpose being to generate randomness.

[20] *SAS* Chapter 5: (*Sat*) Potential Information and Chapter 6; *PGND* Chapter 12.

[21] *SAS* Chapter 6: Information's Infrastructure - Code; also Chapter 20: *DNA*; also Glossary and Index; *PGND* Chapters 12 and 15: Nuclear Super-computing; *A&E* Chapter 2; *SPFP* Lectures 2 & 5.

[22] *SAS* Chap.6; *PGND* Chap. 13 harmonic oscillation, resonance; *SPFP* Lects. 2 & 3.

[23] *SAS* Chapter 6: Machines, also Mind Machines (Computers*): SPFP* Lecture 2; *RSP* Chapter 2.

[24] *SAS* Chapters 5 and 14*; PGND* Chapters 12 and 13*; SPFP* Lectures 2 and 3.

[25] *SAS* Chapter 5: Top Teleology, Chapter 13: Psyche and Psychology, Glossary, Index; *PGND* Chapter 10: Causality and Chapter 13: First State; *SPFP* Lectures 2 and 3.

[26] *SAS* Chapter 0: Delusions, Chapter 3: Hierarchical, Triplex Construction of the Cosmic Pyramid and Chapter 13: Neurological Delusion; *AMA?* Chapters 2 and 3; *PGND* Chapter 13: Neurological Delusion; *SPFP* Lecture 3.

[27] *SAS* Chapter 13: Psyche and Psychology.

[28] *SAS* Chapter 13: The Neurological Delusion.

[29] *SAS* Chapter 13: Consciousness; Slide Set: Consciousness; Slide 22.

[30] *see* Chapter 4 *CUT*; Glossary: unification; also Holy Grails, *GUT*s (Unification Theories) *SAS* Chapter13 and *PGND* Chapter 13.

[31] *SAS* Chapter 14: States of Consciousness.

[32] *SAS figs.* 13.2 and 13.3; *PGND* Chapter 13: Does Brain originate or Mediate? *SPFP* Lecture 3.

[33] *SAS* Chapter 13: Build Yourself a Brain.

[34] *SAS* Chapters 15 and 16; *PGND* Chapter 13: Third Sate of (Sub-) Consciousness; *RSP* Chapter 3.

[35] *SAS* Chapters 16, 17 and 19; *PGND* Chapters 13 and 15; Glossary and Index in these and also in *A&E* and *AMA*; *SPFP* Lecture 3: Third State.

[36] *SAS* Chaps. 15 and 16: Psychosomatic Linkage/ Psychosomasis; *PGND* Chapter 13: Psychosomatic Linkage; *SPFP* Lecture 3.

[37] *SAS fig.* 15.5.

[38] *SAS figs.* 15.3, 4, 5 and Chapter 16: Archetype; *PGND*: Chapter 13; *SPFP* Lecture 3.

[39] *SAS* Chapter 16; *PGND* Chapter 13: *H. electromagneticus*.

[40] *SAS* Chapter 16: Synchromesh 2 - Psychosomasis; *PGND* Chapter 13 last three sections; *see* also Glossary: resonance and indices: resonance, harmonic oscillation and vibration; *SPFP* Lecture 3; *RSP* Chapter 3.

[41] *SAS* Chap. 17; *A&E* Chap. 17; *AMA?* Chap. 24; *PGND figs.* 13.7 & 15.1; *SPFP* Lecture 3; *see* also Glossary: *prana*.

[42] *see* also *PGND* Chapter 13: Triplex Psychology Summarised.

[43] *SAS* Chapter 8: (Physical) Holy Grails; *PGND* Chapter 13: First State; *SPFP* Lectures 3 and 4. Once the penny drops that Natural Dialectic is a construction of Cosmic Connections whose aim is to build towards a Holy Grail of Unification, one that includes all aspects, physical *and* metaphysical, of cosmos, then the term Cosmic Connections acquires fresh resonance; and, with such resonance, social,

political and psychological impact (see Chapter 6 and also *SAS* Book 4 and *PGND* Chapter 16).

⁴⁴ *see* Glossary: also for matter-in-practice.

⁴⁵ *SAS* Chapter 7: Lady Luck and Lord Deliberate *passim*; *PGND* and *SPFP see* Glossaries: chaos, Indices: chance.

⁴⁶ *see* also Appendix 1.

⁴⁷ https://evolutionnews.org/2010/04/roger_penrose_on_cosmic_finetu/

⁴⁸ *SAS* Chapter 9.

⁴⁹ *SAS* Chapter 10.

⁵⁰ *SAS* Chapters 11 and 12; *SPFP* Lectures 2 and 3.

⁵¹ *SAS* Chapters 11and 12; *PGND* Chapter 8; *A&E* Chapter 16; *SPFP* Lecture 4; *RSP* Chapter 4..

⁵² https://en.wikiquote.org/wiki/Max_Planck

⁵³ *SAS* Chapters 19 and 20; *SAS* Glossary and Index: code; *A&E* Chapter 4: Genes and Genesis; *AMA?* Chapter 22; *PGND* Chapter 15; *SPFP* Lecture 5.

⁵⁴ *SAS* Chapter 23: Super-codes and Adaptive Potential

⁵⁵ *see* Glossary: **conceptual biology** and archetype.

⁵⁶ *SAS* Chapter 20: Alchemy - 20+ cumulative reasons why chemical evolution is a no-go.

⁵⁷ *SAS* Chapters 20 and 21 *passim*; *PGND* Chapter 15: Chemical Evolution?; *A&E* Chapter 9; *SPFP* Lecture 5; *RSP* Chapter5.

⁵⁸ *GTE* (general theory of evolution) hypothesizes unlimited plasticity of transformation; *see* Glossary: macroevolution; also *SAS* Chaps 22 and 23 & *PGND* Chap 15.

⁵⁹ The Encyclopedia of *DNA* Elements (ENCODE) is a public research project launched by the US National Genome Research Institute in 2003 and continuing apace. Its aim is to identify all functional elements in the human genome.

⁶⁰ *SAS* Chapters 5, 6 and 19; *PGND* Chapter 15; *SPFP* Lecture 2.

⁶¹ *see* Glossary; also *SAS* Chapter 19: The Central Executive is Homeostasis; Index: balance and equilibrium; *see* also *PGND* Chapter 15; *SPFP* Lecture 5.

⁶² *SAS* Chaps 5, 6, 14, 17, 19, 23, 25; *PGND* Chaps 11-13.

⁶³ *SAS* Chap. 21: Energy Metabolism Perchance?; *A&E* Chap. 10: Life's Engine; *SPFP* Lecture 5; since chemosynthetic systems are different yet as complex as photosynthetic ideas that convergence, coevolution or that the latter might have taken over from the former are, since they do not explain system origin, fanciful.

⁶⁴ Glossary: *ATP*.

⁶⁵ *SAS* Chapter 19: Nuclear Super-computing and Conceptual Biology; *SPFP* Lecture 5; also check Indices: switch.

⁶⁶ *SAS* Chapter 23 and Glossaries.

⁶⁷ *SAS* Chapter 22: *A&E* Chapter 3: Hierarchy; *SPFP* Lectures 2-5.

⁶⁸ *SAS* Chapter 22: The Origin of Secies; *AMA?* Chapter 9: When is a Man a Man?; *SPFP* Lecture 5.

⁶⁹ *SAS* Chapters 20 and 21; *PGND* Chapter 15: Chemical Evolution?; *A&E* Chapters 4, 8, 9 and 10.

[70] *SAS* Chap. 22: Tree of Life; also 'genetic discordance' and 'anomalous gene trees'; *SPFP* Lecture 5.

[71] *SAS* Chap 22: Fossils; *A&E* Chapters 13 and 16; also Meyer: Darwin's Doubt..

[72] *SAS* Chap. 22: The Origin of Type; *see* also bio-calssification and archetype.

[73] *SAS* Chap. 20 *passim* (over 20 cumulative reasons).

[74] *SAS* Chap. 22: The Editor; *A&E* Chap. 5: Sports, Survival and the Hone; *SPFP* Lecture 5; *RSP* Chapter 5.

[75] *SAS* Chapter 23: The Creator; *A&E* Chapter 5; *PGND* Chapter 15: What's the Problem?; *SPFP* Lecture 5; *RSP* Chapter 5.

[76] *SAS* Chapters 21, 23 and 25; *A&E* Chapters 2 and 9; *PGND* Chapter 15; *SPFP* Lectures 2 and 5 esp. concerning development; *RSP* Chapter 5.

[77] In the 'Origin of Bio-information and the higher taxonomic categories' (Proceedings of the Biological Society of Washington 4-8-2004) S. Meyer argues that no mechanistic theory can account for the amount of information needed to build novel forms, systems or organs of life.

[78] *SAS* Chapters 24 and 25; *A&E* Chapters 7 and 8 for more detail as regards Reproduction and Reproductive Archetype; *SPFP* Lecture 5; *RSP* Chapter 5.

[79] *SAS* Chapter 24; *A&E*: Chapters 7 and 12; *SPFP* Lecture 5.

[80] *SAS* Chapter 24; *A&E* Chapters 7 and 13.

[81] *SAS* Chapter 25 *passim*; *A&E* Chapter 8; *SPFP* Lecture 5; *RSP* Chapter 5.

[82] *see* Chapters 3 and 5, Appendix 3 and Glossary; also *SAS* Chapters 16 and 17.

[83] *SAS* Chapter 26; *SPFP* Lecture 6.

[84] Absolute Community of Essence and Existence: *see* also *SAS* Chapter 4 and *PGND* Chapter 16: Truth, Appearance and Reality.

[85] *SAS* Chapter 3: Hierarchical Perspective, *fig.* 3.5 Cosmological Bearings; also *PGND fig.* 7.4; *SPFP* Lecture 2.

[86] Ecology: *SAS* Chapter 26.

[87] *SAS* Chapter 26; *RSP* Chapter 6.

[88] *SAS* Chapter 4: Is there an Absolute Morality?

[89] *SAS* Chapters 4 and 26; *PGND* Chapter 26; *SPFP* Lecture 6.

[90] 'Religion' is derived from the Latin 'to bind back'; 'yoga', equally, means binding (to a discipline); Nuclear Religion; *see SAS* Chapter 5 and 26: Top Teleology; *PGND* Chapter 13 Transcendence; *SPFP* Lectures 2, 3 and 6.

[91] Individual Association: *SAS* Chapters 4, 26 and 27; *PGND* 16: Science to Conscience and Is There an Absolute Morality?; *SPFP* Lecture 6; *RSP* Chapter 6.

[92] if indeed that ever happened; *see AMA? passim*.

[93] *see* Glossary: Vitruvian Man.

[94] *SAS* Chapter 4 and *PGND* Chapter 16: Truth, Appearance and Reality; some might treat Essential Characters (Primary Dialectic, right-hand column) as facets of One Natural Creator, a Truth and Paramount Reality labelled word-wise 'God'.

[95] *SAS*: Chapter 1 Natural Dialectic's ABC/ Index: stack and dialectical operator; *PGND* 5 & 6; also Index: stack and dialectical factor; *SPFP* Lect. 1: basic grammar.

[96] *PGND* Chapter 8: Essence; Chapter 10: existence; *SPFP* Lecture 2.

[97] *SAS* Chapters 8: Principles of a Unified Theory of Matter; and 9: Nothing.

[98] *SAS* Chapter 8: Infinity and Index; an infinity of existential units can be mathematically defined; The Unity of Infinite Essence cannot.

[99] *SAS* Chaps. 2, 4, 8: appearances, relativities and lesser, seeming elements; *PGND* Chap. 8: essential characteristics in existence; *SPFP* Lecture 1: limitations of essential characters.

[100] *SAS:* Chapters 8, 9 and 10.

[101] *SAS fig.* 2.4; *PGND* Chap. 8: positive and negative zeroes; Chap. 10 inversion; *SPFP* Lecture 1: inversion/ reflective asymmetry.

[102] Both with respect to metabolic dynamics and (Chapter 25: *passim*) development.

[103] *see* also Science and the Soul (*SAS*) Chapter 16.

Index

functional logic 39

G

gene

allele...93

micro-hierarchy/ genetic regulation/
expression..89

pool 93

genetic linguistic hierarchy

genome/ encyclopaedia/ book of life.......89

gravity/ anti-levity

dialectical sense

subjective - informative loss > noise/
nonsense/ non-conscious state.......*see*
informative entropy/ psycho-logic/
randomness/ oblivion

subjective > inc. suffering/ igorance/
unhappiness/ pain...........................110

scientific (and dialectical) sense

mutual attraction of physical bodies/
gravitation 23

GTE (general theory of evolution). 82, 83,
90, 97, *see* also Index; evolutionary
theory and Glossary; macroevolution

GUT

grand unified theory of physics 67

H

H. archetypalis. *see* also man/ archetype/
first cause physical/ typical mnemone,
see also typical mnemone human
psychosomatic form/ mind-side.........54, 58

H. electromagneticus .... *see* also man/
electrical charge

electrodynamics 57, 58, 137

non-conscious bioelectrical aspect 57,
134

psychosomatic form/ body-side 135

quantum physique/ quantum man 137,
see also matter-in-principle

H. sapiens........... *see* also man/ bio-logic

non-conscious expression of bio-logic ...*see*
also anatomy/ bio-logic/ bio-logical
development

H. sapiens-physical, non-conscious
expression of bio-logic 134

harmonic oscillation 58

harmony .... *See* music/ health/ resonance

health...38, 60

ecological...108, 109

harmony-in-action....................................38

heaven-on-earth/ utopia.....................110

hermaphrodite ....*see* sex/ archetype/
neutrality/ reproductive polarity

hierarchy

bio-logical hierarchies......................87, 100

computational hierarchies.... 88, 99, *see* also
computation/ informative hierarchy/
code

informative > energetic act of creation ...24,
88, 136, 137, *see* also conceptual
development/ creativity/ information

symbolic 44

hippocampus 53

HIV..96

holism 10

*both immaterial-material/ psychological-
physical/ informative-energetic
components of cosmos included* 7, 8, 10,
20, 111

Holy Grail 66

homeostasis .86, see also health/ balance

bio-systematic..84

triplex mechanism/ sensor/ regulator/
effector 84

homology

archetypal routine/ bio-modular
programming...89

Hox gene cluster 101, 102

I

ignorance *opp.* knowledge...................51

Illumination .....33, 38, 112, 114, *see* also
Consciousness

immorality *opp.* morality > darkness ..*see*
morality

immortality

genetic...87

Uncaused Cause/ Pure Deathless Life 27

impotence

final cosmic phase/ non-conscious
energetic matter....................................46

incarnation

biological organism...................................53

Infinity...*see* Absolute/ Natural Essence/
Trancendence

information

175

T

The author has recently written a few more books (available from Amazon, Foyles, Waterstones, Barnes & Noble etc. and see website addresses on p.2):

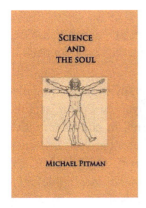
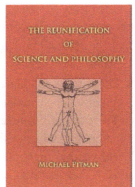
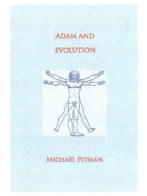

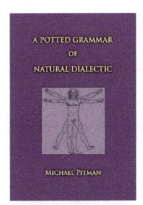
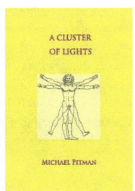
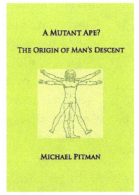

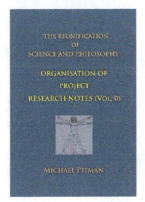
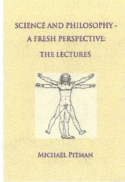
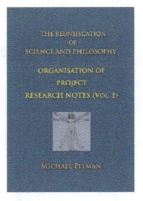

Contents illustration credit: wikiHow creative commons (BY-NC-SA) license; wikipedia, wikimedia commons.

Answers

Chapter 1

1. Materialism believes everything is composed of non-conscious energy and particles; holism allows material *and* immaterial elements.

2. No. Holistic Natural Dialectic is fundamentally separate. Its polar root includes, as a key component, metaphysic. It thus rejects the basic theoretical axiom of scientific materialism and its narrative that the events of cosmos are only physically derived. Instead it proposes, with different cosmic outcome, both metaphysical *and* physical considerations. It thus amounts either to 'negligible' non-science or a systematically worked 'paradigm shift'.

3. Information, mind, consciousness, thought and so on.

4. A theory of opposites; a binary system that reflects the oscillatory nature and basic order of cosmos.

5. That non-conscious molecules gradually aggregated by chance into simple and, over a long period of time, complex organisms by chance mutation and natural selection (Chapter 5).

6. A pair of scales, concentric circles and a stepped pyramid called a ziggurat.

7. At the base.

8. At the periphery.

9. ↓ sink source flow ↑

10. Source; in its materialising direction Natural Dialectic always runs from source through action to sink.

11. Scale: pivot; concentric circles: centre; ziggurat (Mount Universe): peak.

12. ↓ finish   start   action   ↑   or, as in chemistry,
    ↓ completion precondition reaction ↑

13. Metaphysical.

14. Three abstractions or qualities whose tendencies are expressed, in relative degree) by every object and event (physical and metaphysical) in cosmos.

15. (↑) up, (↓) down and balance or, for example, (↓) yin, Tao and (↑) yang.

16. As projection from source of light or Trans-existential Light. The conscio-material spectrum (running outwards from source) or materio-conscious spectrum (inwards from periphery) are simply anti-parallel ways of understanding the single projection. Is creation really a projection by the Immaterial Absolute like that of a projection from source of a physical absolute, the finest matter known as light?

17. The spectrum is also known as an existential dipole composed of information and energy.

18. Existence is motion and forms; it is ceaseless change; and it involves both mind and matter.

19. Essence (Latin: being); essence *is* without predicate; it is no-thing as opposed to any local thing, any temporal phenomenon; and it is the Unconditioned, Unbounded (Infinite) Singularity from which the duality of existence is derived.

20. Of Essential Qualities (set in the right-hand column) against (on the left-hand column) their existential opposites e.g. Balance set against imbalance, Potential against expression or Neutrality against polarity.

21. Of vectored opposites, as when Neutrality is polarised into (↑) positive and negative, (↑) stimulus/ (↓) drag or (↑) up/ ↓) down (as opposed to balance).

22. Because, in any local/ temporary case, balance is locally represented by degrees of *more or less* balance/ imbalance; it is, set against the Absolute, Essential Criterion of Perfect Balance, an existential form of *apparent* balance.

23. Straight-line, circular or wave-form (vibratory).

24. Zig-zag, ellipse, irregular straight line or cyclical motion.

25. A dot.

26. Relative.

27. Aspects of Essence.

28. Aspects of ever-changing existence.

29. Polar aspects of the left-hand Primary character; for example Neutrality is opposed by polarity; the character 'polarity' is then split, in secondary dialectic, to (↑) positive and (↓) negative. This triplex order is universal.

30. Potential (pre-active), active and passive (post-active/ impotent/ fixed) divisions.

31. For example, any one of the questions framed near the start of this chapter. The 'philosophical machinery' of Natural Dialectic employs a simple, stepwise protocol following the three cosmic fundamental (e.g. start, action, end or source, flow, sink). Such algorithm amounts, as we'll see, to a 'core formula' or 'framework of common sense' within which to approach questions involving the nature of cosmos or our lives in it.

Chapter 2

1. Cosmogony (the origin of physicality), the nature and origin of consciousness and the origin of codified life-forms.
2. Observation from the perspective of an inventor or expert who knows his subject thoroughly and works from principle to detail; it is also the order of a creation. For *top-down* programming see Index: computation.
3. That used by a learner who works from detail to principle.
4. Abductive reason is applied in historical or futuristic cases; its speculation involves, by the elimination of competing possibilities, best guessing as to the cause of any particular instance.
5. History is untestable unless it can be accurately reproduced in experiments today. Thus theories of origin are not strictly scientific; and, as Sherlock Holmes, was well aware, 'unseen variables' may crop up - ones which could change the whole direction of the chase. There are two possible interpretations of historical nature and, at root, origins. Did not an atheistic revolution guillotine opponents? Materialism, truncating the universe, works without a head. Holism, on the other hand, accepts that tracking down the origin of systems and of integrated, codified and cooperative mechanisms, invariably confronts a mind at root. Thus, while one accuses the other of 'gods-of-the-gaps' the other points to billions of unseen 'mutations-of-the-gaps' that sustain the evolutionary version of our past! Where does truth go begging?
6. The basic binary duality is a dipole consisting of the Singularity of Infinite Essence and binary existence.
7. Energy and information.
8. Transcendent, subjective and objective; or potential, mind and matter.
9. Archetypal, quantum and bulk material levels.

10. Vertical causation. Such causation includes the metaphysical dimension. Its line of operation runs from idea though planning to material effect; any act of creativity is hierarchically expressed through phases into material arrangement - even if the end-product is simply nervous excitation without effect beyond the body. See also answer 15: 'finger-waggle'.

11. An implicit (unseen) first cause; the source of issues as explicit effects. There are two kinds of universal archetype - higher, active, conscious and lower, passive, non-conscious. The latter is also called causal or potential matter.

12. Top transcendence and base subtendence.

13. The extreme subtendency of a non-conscious singularity (say, black-hole or pure space) and extreme transcendence of conscious singularity (pure concentrate of consciousness).

14. Both.

15. It is the order of creativity; it includes prior psychological and subsequent physical phases.

16. Check the index for 'finger-waggle'. It is the way idea, desire or plan is physically expressed. These creators are immaterial, in mind; the product (say, an action or a mechanism) is physical. Such line of operation illustrates the order of an act of creation. In this case, running from metaphysical to physical arenas, the 'finger-waggle' is the end-product. So is the chair you are sitting on and all parts of the city you live in. They are all end-products having mind-in-them; not active, brainful mind but the shape of thought that dynamically informed them and without which they would have no being. It is important to realise mind all about you. It is natural and, though, invisible to the instruments of scientific analysis, inevitably present. Do we not ascribe works to their authors even though the passive carriers of their instruction, books or artefacts, have no mind of their own. No analysis is complete without nod to its implicit but physically absent creator.

17. 'Information is information, neither matter nor energy.'

18. In an individual sense it is the precondition or context in which an idea may arise (see The Order of an Act of Creation); in a universal sense is either conscious First Cause (Psychological Archetype or Source of Mind) or first cause physical (archetype we call typical mnemone; see Appendix 3). The goal of sages is to unite with

Logos, this Source of Mind or Living Self. This Self is not our mundane body-self nor ever-changing mind. It transcends both 'coverings'. And Living Self is, paradoxically, at the same time both individual and cosmic. Thus the injunction "Know Thyself".

19. Active information is conscious mind's creative field; informant mind generates purpose, logic, images and code. It is teleological.

20. Any informed item or expression of active mind. Memory (subconscious mind) is passively informed; so are all non-conscious physical forms and events. Passive information is aimless; it is non-teleological.

21. Shannon's theory of information is mathematical. It is statistical and deals with the expressed form of code by counting and comparing its digits whether these be in alphabetical, numerical or electronic form. Therefore, it takes no account of vertical causation, that is, the mind that produced the passively informed, physically reflected version of code. Importantly, it takes no account of consciousness, mental process, purpose, meaning or emotion - all critical, teleological elements of the 'active informant' we call life.

22. Purposive complexity is the product of mind (e.g. machines, codes etc.); non-purposive complexity is the aimless aggregate of non-conscious forces.

23. Mind (the immaterial informant with its ideas, purposes and plans). Never, in principle, throughout the whole of space and time could natural forces (composed of non-conscious matter and energy) signal a meaningful, purposive message or create a mechanism. This is not how material nature works - unless you are forced against the evidence and grain to claim that theory trumps reality and senseless 'natural force' of evolution coded, messaged and constructed mechanisms by the million!

24. Codes, plans, mechanisms and machines are irreducible to physics and chemistry because, just like the chair you are sitting in, they involve vertical causation. In this respect there is, inseparably, mind in them. Such causation is not reflex; it is hierarchical (check Chapter 1: "finger-waggle").The question is, is *all* matter minded? For further information see Glossary: 'code' and 'machine'.

25. Although conceived by mind computers are, like brain, passive calculators. They are programmed to act reflexively to instructions. However, since much psychological activity is also the product of

program (instinct and habit) computers mimic mind. But, composed solely of non-conscious materials, are they conscious? Can they experience anything? Ever?

Chapter 3

1. Soul, mind, body.
2. ↓ non-conscious body   Soul   conscious mind ↑
3. An illusory, perhaps even deluded, imagination of mind; a figment of unreality. If 'common sense' is what your senses can perceive, then 'soul' is nonsense. If, however, 'common sense' pragmatically includes the immaterial elements of mind, experience and consciousness then, at the heart of these, 'soul' makes sense.
4. Alive, subjective, natural and paradoxical; Most Natural, at once both everything and nothing, soul is nature's seed. It can be experienced as one. One is not two so that in communion your *ego* will be wholly lost. And, since imagination involves forms of thought, soul cannot be imagined. Yet, nearer than breathing, soul is your life's inseparable centre, core, the Self of 'Know Thyself'.
5. It certainly treats the physical, neurological aspect of brain; it also deals in multiple, often complex theories of mind's *ego* (the body-attached *self* of life on earth); however, the nature of dormant mind (instinct and the subconscious) is unclear; and full psychiatry (from the Greek 'soul-healing') remains the preserve of 'unprofessional pseudo-scientists' such as saints. Who is missing the centrality?
6. A brain with associated nervous system, sense and motor organs and, indeed, a whole attached body. Thought is thought to somehow reside in nerves; conscious experience is known as being caused, somehow, from a complexity of inanimate electronic charge.
7. That 'mind is meat'; that, although composed of non-conscious chemicals, nerves, especially in the context of brain, somehow 'exude' experience; in short, the delusion is that brain *is* mind.
8. No. Like the rest of the biochemical body, it is *per se* non-conscious; in holistic terms it is, like the control panel in an aircraft allowing a pilot to fly, a mediator through which sensory impressions flow in and motor actions respond to the pilot's purposes; if body is the plane, brain is designed to automate many processes yet allow pilot override on the flight through a biological life; in short, brain knows no more than a brick. Mind is *not* brain.
9. Depending on your perspective at the third eye, a metaphysical location between and just above and behind the two physical eyes

or at the physical, cerebral cortex. Cerebral cortex is long known as the organ of thought (you may touch your forehead when concentrating) and consciousness. This does not necessarily mean that it generates subjective experience, as neuroscience presumes. If brain is a mediator it simply means that this is the *PSI* area of 'impact' between conscious, subjective mind and objective, physical context. All areas of computer-like brain respond to thought just as they respond to incoming sense impressions and issue outgoing motor orders. This does not detract from scientific study of the physical side of interface; it simply means that such study will, even in perfection, not yield the full story - because brain is not mind.

10. The level less interested in mundane detail than the principles that substantiate our being; scientists and philosophers both employ higher mind.

11. 'Horizontal' causation involves only the physical tier of creation with reflex (aimless) knock-on effects according to 'natural law'. 'Vertical' causation also involves the metaphysical tier of mind.

12. No. Reflex reactions occur 'horizontally'. But if mind is involved we call causation 'vertical' (cf. Index: finger-waggle). We say that thought precedes an action. If you equate the word 'natural' with 'material' then immaterial mind has the potential to rearrange material conditions in an 'unnatural' way Its subjective potential works, through body, to rearrange objective impotence, that is, material things.

13. Of course. Reflex or automatic action and reaction are the nature of physicality. The question then is "Whence did/ does physicality originate?"

14. Potential precedes particular expression; in holistic terms, it is the implicit possibility that could allow a particular, explicit event to occur; like rules (mentally devised) allow myriad particular expressions during actual games, so the product of unseen potential is local and specific action/ change/ transformation.

15. An archetype is the pre-active potential for the cosmic level below. It is the first cause of what follows. Thus mind's source is consciousness; its archetype is conscious. Matter's potential is archetypal memory in universal mind; a memory is a subconscious factor; therefore, the archetype of non-conscious matter is unconscious.

16. Two. Of immaterial psychology and material physic (which latter we tend, deficiently, to think of as the whole natural world).

17. Super-conscious; consciously awake (focused, poised); consciously awake (restricted to sense-based responses, egotistic); sub-conscious/ dormant (dreaming); sub-conscious/ dormant (in deep sleep or coma).

18. A mask or *persona* veiling the true self. The question then arises 'Who am I really?' In this case we construe *ego* as our notion of identity, that is, who we think we are in the context of past and present circumstances while inhabiting a specific body. It is a 'thought-box' that comprises self. Personal *egos* are all different but yours, unlike a cat's, bee's or newt's, is of human kind. If your self becomes involved with the inflation of natural urges it becomes inflamed in the form of various passions such as lust, anger, greed, attachment and arrogance. Balancing the ego-self (me) against the influence of such 'evils' is a major component of life in society of any kind (see Chapter 6).

19. A memory. If mind is analogised with matter, its three main conditions are gas (super-conscious freedom), liquid (the normal conscious state we are in now) and solid (its subconscious databank composed of files called memories). In spectral terms, we say ultra, visible and infra wavebands.

20. A databank of tagged files called memories.

21. Body's informant is placed at the 'head' of body. The front cortex is involved consciousness; central is the linkage between electrical or hormonal information systems; around the centre are aligned somato-sensory and motor inversions (from foot at top round to mouth at base) which (also by inversion) together serve the nerves and muscles of the left-hand (from right hemisphere) and right-hand (from left hemisphere) sides of the body.

22. No. For reasons of code.

23. Where fixed, subconscious mind meets active matter (matter-in-principle or the quantum band of events). We call this junction the psychosomatic interface or *PSI*.

24. An outer shell, called your physical body, or biological phenotype, encases several 'wireless' bodies. These are, moving inwards as if a Russian doll, the unconscious psychosomatic shells composed of quantum patterns of light and charge (*H. electromagneticus*) and

archetypal or potential matter (sometimes called the etheric body); next up is conscious mind, the body of your instant perceptions; above this range various bands of higher consciousness, not all accessible to normal human experience; at your core is the concentrate of consciousness called *H. spiritualis* or soul. In this case soul is the heart of life.

25. Two: *H. electromagneticus* and phenotypic *H. sapiens*. The former is treated by molecular and quantum biologies, the latter by classical study of form and function

26. The junction between conscious and subconscious mind.

27. The junction between subconscious and body; the psychosomatic interface (*PSI*).

28. A division of memory.

29. Personal and typical. Personal mnemone is your own databank; typical mnemone is the archetypal memory for a particular type of body (not necessarily in alignment with the unnatural, humanly conceived classification but defined, according to that classification, at the level of family, order or class). Personal memory maybe likened to a particular program broadcast on the back of its own kind's typical channel e.g. your individual databank recorded on the human channel.

30. Signal translation, instinct and morphogene; signal translation translates nervous blips to conscious experience; instinct is the framework of a creature's psychological behaviour; and morphogene is the tympanum of its physical shape(s), the 'chip' mediating archetypal and quantum instructions.

31. In every cell in every body. It is psychological '*DNA*'.

32. The study of waveform with especial interest in the relationship of sound waves to physical form. Study of the vibratory transfer of energy especially at controlled, specific frequencies. Since code is transmitted by light or sound (not least the orderly vibrations of your own voice) the transmission of codified energy is apposite to the two-way traffic of psychosomatic signals.

33. Attunement. Resonant association, as in cymatics, between signals from typical mnemone and quantum antennae inherent in configurations of electronic charge.

34. Typical memory is a read-only cache laid down in universal mind. Not so personal memories. Incoming (sense) data impacts cerebral

nerves and creates localised nervous patterns linked, by signal translation, to conscious experience and back, in the other direction, to outgoing motor response. A read-write data file needs be tagged for future retrieval and the suggested label is created, automatically, by the nervous configurations associated with any particular event. Whether this localised 'address key' or 'pin-code' is logged with a central reference library (at, say, the amygdala and/ or hippocampus) is unknown. However, if some key configuration is triggered by a later, similar train of thought, emotion or sensation, the memory (that is, the image of the original experience) will be automatically retrieved. In the reverse direction, a memory retrieved into conscious mind can also trigger responses according, by computation, to the cerebral locations of the nervous configurations that originally responded to the original perception. In other words, a memory retrieved to consciousness can back-track to trigger the physical associations that accompanied its creation. The instrument of resonant association at the key localities would reiterate the same or similar psychosomatic algorithm as before. Thus, while the bank of personal files in metaphysical memory would remain intact, physical degeneration of nerves could impair the process of tagging by address keys and subsequent physical responses by way of such faulty keys. Now, what do you think? It's time for creative (as opposed to critical) ideas of your own.

35. By the hierarchical standards of Natural Dialectic, they are superbly logical. They reflect bands on the conscio-material spectrum. In fact, all organisms display their own 'spectrographic' fingerprints but, for better or for worse, the human sequence as shown in the illustration 'Informed Man' indicates the most complete reflection of lines and bands.

Chapter 4

1. Potential matter (archetype), matter-in-principle (quantum particles and forces), matter-in-practice (aggregations of mass which compose bulk matter).

2. ↓ passive   potential   active ↑   to which you might add, say,

fixed energy/ (mass)	archetype	quantum behaviours
contraction	latency	radiance

gravity	balance	levity
drag	equilibration	stimulus

3. Yes. We all can. The scientist is using sense perception and it technological extensions to better understand the 'outside' or world of non-conscious bodies even when it comes to the study of biological forms. The mystic turns 'inward' toward the source of subjective, immaterial mind. All humans employ each anti-parallel direction of focus on a moment-to-moment basis but do not systematically sharpen their attention in the way of scientist or meditator.

4. To find ultimate truth; to achieve utmost simplification, unification or basis. The current holy grail of physics is called a grand unification theory (*GUT*) whereby quantum and classical physics can be united; this has not happened, despite string and other speculations, because a satisfactory theory of quantum gravity has not yet been achieved. The nature of this grail may change if 'hidden variables' or other factors, such as 'dark energy' or 'lambda force' have to be included. Holistic theory assumes a 'hidden invariable' in the form of immaterial archetype.

5. It is always reflex and incapable of design. Nature could not, in a billion years, chance to create a cup of sweet tea let alone drink it.

6. Potential matter. Archetype. Electrical potentials, the potential of position or of balance to tip one way or another, gravitational potential or other field-derived properties of physical science are not what holism means by the term, although it is in common with the notion of unrealised abilities.

7. Again the notion is broader than the specific meaning of mutual mass attraction. It applies to any contractive, compressive, inertial or materialising tendency; as a quality of the ($\downarrow$) *tam*, downward fundamental its vector 'bears down' and 'outwards' towards the base of cosmos. Its opposite is levity which implies liberation as opposed to capture, free energy and stimulus for change.

8. Metaphysic. Mathematics deals in symbols. These are immaterial. They are a property of metaphysical mind's creativity and not material objects or events.

9. Nothing. Can anything come from nothing? Or was the absence of physicality not metaphysically void? Either way, whether one proposes eternal matter or eternal consciousness, we are proposing

the greatest miracle. This world is either a miraculous projection from a metaphysical source or the energy that underwrites is eternal! The question is what you believe. In which case have you invested faith?

10. Because the initial conditions of cosmos appear to have included a 'pin-code' of about thirty very precise, crucial and cooperative settings called constants.

11. The 'steady-state' assumption of eternal energy/ matter is out of favour. It is believed cosmos began suddenly, nowhere, out of nothing for no reason in a mysterious projection called the 'big bang'. This idea has problems although for nearly a century objections have been met by ideas like, for example, 'inflation'. Nowadays it is popular to believe that a multiverse might strike up enough tries to 'crack the code' for our particular universe. Chance could be restored as its creator. There is absolutely no proof whatsoever for this huge, speculative fudge, a very long-odds bet which amounts, simply, to materialism's cry for any contrivance to try and eliminate pre-conditional plan!

12. Creation of a single cell (or in the sexual case two) containing the information to create a mature body is analogous to the holistic view of a pre-codified, orderly-linked and working body of cosmos.

13. Because they form the inner structural basis from which the outer, sensible details of the world are derived. Quantum is sometimes called 'subtle' while the 'iceberg aggregates' called complex, massive things that float in its sea of energy are called 'gross'.

14. What is chance? Is it complete randomness? Or the probable odds that something might turn out in a certain way as opposed to a range of other possibilities ranging from certain to impossible? The established view of most quantum physicists is that, since you can never capture all data about a particle you can never know exactly what to expect of it. Such indeterminism holds that cosmos slides across a substrate of probability so that, at root, nothing is certain, all is a matter of chance. While this may be true of single particle behaviour their aggregation into fixed atomic, molecular and higher structures whittles the notion of chance to virtual, perhaps absolute, impossibility. Such determinism at physical level is corroborated by invariant 'rules' that seem to govern physical behaviour. In order to scotch ideas of determinism, invariance and there for plan some

physicists are attempting to show that 'the laws of nature' are themselves flexible. Can any new ones be evolved?

15. An 'alphabet' of specific vibrations whose grammatical construction the forces spell out, instruct or control. These vibrations are also analogised to musical notes and, in this case, creation thought of as an energetic vibratory scale in which its various bands are 'registers' or 'octaves'. So, within the rules, individual possibilities appear as endlessly varied yet orderly as speech, musical composition or other creative transformations.

Chapter 5

1. Elucidation of the structure of DNA.

2. Information in the form of code. This code is exceptional in that it translates its symbols into what these symbols represent. In the way of a book being able to physically realise its story DNA code controls the construction of what it codes for.

3. It is metaphysical. Information is immaterial, that is, metaphysical. Thus the basis of biology is, although expressed from a chemical alphabet, actually metaphysical.

4. No. They are chalk and cheese. One involves meaning, purpose and deliberate organisation; the other, an apparent property of non-conscious reflex, aimlessness, never does.

5. There is hope, guess, theory and theory-driven laboratory experiment (the intelligence of scientists trying to make such a macromolecule) but no proof at all. Intense scientific effort has failed to produce a single, sufficient bio-polynucleotide. In nature entropy would have degraded it anyway in hours or at most days. To 'work', DNA and RNA both need a host of accurately shaped concomitant biochemicals. Why should atoms or complex molecules built of them 'want' to collect in a micro-space in order to create, against the natural force of entropic breakdown, survival? Cellular machinery, like any machinery, is an entirely different matter. But, as we saw in Chapter 2, machines need informative design. Machines and natural, non-purposive structures are chalk and cheese.

6. No. The ideas behind chemical evolution are irrational. The reverse of reason, anti-reason, is chance. The mindless incoherence of random events cannot develop codes, mechanisms or the integrated

systems of any purposeful machine. It is therefore highly irrational, even if it suits materialism's theory, to suppose that aimless events can generate a reproductive bio-machine called a cell.

7. No. The Second Law of Thermodynamics involves a property called entropy. This means that a system will tend to break down over time unless energy is input to counter the tendency. Thus, since an input of raw energy such as sunlight is liable to destroy rather than construct any complex system, eo that even a single bio-molecule (such as a polyglyceride, polynucleotide or protein) occurring accidentally is vanishingly small and liable to be broken down rapidly. All bio-molecules are, in fact, the direct product of pre-existent code. Thus the Second Law, entropy and the fact of prior codification for replication appear in concert to work very strongly against the notion of bio-evolution.

8. *E cellula cellula* - from a cell a cell, from parent offspring. No exception to this fact has actually been found; Pasteur found proof of it. Yet abiogenesis, a crucial requirement of materialism's creation story, absolutely, and entirely hypothetically, contradicts it.

9. Darwin saw a species as we do today, a classification subject to variation.

10. Darwin's guess was to assume unlimited plasticity of species; in other words he made the wrong guess because he was not in possession of the molecular facts. He chose to guess unlimited plasticity (macroevolution or the general theory of evolution, *GTE*) whereas the basis of biology is in fact information in the codified form of highly complex programs, the alphabet of whose quaternary expression is represented by DNA bases. In short, he should have concentrated not on an origin of species but the origin of information.

11. Time is not the issue. Time, chance and natural forces are not creators of machinery. Mind is.

12. Codes and signals. The boss of bio-process is plan. DNA is, *per se*, lifeless. Passive. It is simply, like the program fed into computer, a component in the co-operations it serves. A non-conscious machine is absolutely not its own conscious engineer

13. They anticipate; they guide and control; they involve goals. The hierarchies, algorithms and the chemical agents of their pathways are instructed to produce specific end-products.

14. Because a signal instructs one or a series of controlled, dependent intermediates to execute its plan (or algorithm) and obtain the required goal.

15. No. Only intelligence, not the wind and rain of natural forces, can produce a *top-down* system of instructions in order to achieve a goal. In biology, for example, the top level of any outcome is code controlling precise metabolism that in turn supports specific functions and structures of a body; development from egg to adult body is another hierarchical process.

16. Dynamic equilibrium is a state of balance maintained between continuing processes. In the bio-case it involves the maintenance of equilibrium between processes of molecular breakdown and reconstitution of an inherently unstable constitution. The ups and downs of such a process, when regulated within prescribed margins, can be drawn as a waveform. A wave is simply a circle extended through time. So the process is vibratory and cyclical around a preset norm.

17. Homeostasis.

18. No. Homeostasis is a triplex process. Its cybernetic feedback system (an automation only developed recently by human intelligence) is a mechanism comprising three sub-mechanisms, each with its co-operative component part(s). The codified process needs prior establishment of the norm to be maintained in coordination with all the other contingent norms of linked processes; its parts include an organ (maybe a molecule) called a sensor to signal condition; a processor to match the signal information against norm; and an executive algorithm to ensure the norm, operationally attuned wth organs of the body as a whole, is maintained. The critical cybernetic process cannot start without all three component units in place.

19. The previous questions 2-18 have indicated the outline of an answer here. The basis of biology is information and life-forms are, without exception, codified. Chapter 2 demonstrates why code never arises by the fortuitous aggregation of non-conscious particles and play of forces such as occur in, for example, sunlight, wind, rain, seas or soils. The irrationality of the notion of chemical evolution is further elaborated in over 25 cumulative steps in Science and the Soul (*SAS* Chapters 20 and 21: see *Connections* for acronyms) and *A&E* Chapter 4. The demonstrable absurdity of Darwin's guess (the

theory of evolution or unlimited plasticity) is in turn elaborated in the One World Set, *SAS* and othe books. **Neither mythical nor religious, such elaborations are solely logical and scientific.**

20. As usual, the cosmic fundamentals. Potential, action, end-product. Information, in the form of a purpose or plan, always precedes its physical expression. The idea to do something precedes the action's execution to achieve the desired result.

21. Therefore write:

↓ end-product	potential	steps in process ↑
passive structure	information	energy-for-action
component parts	architectural plan	metabolism
structure	objective/ goal	function

22. The *DNA*-translated-into-protein (transcription-translation) system, metabolic feedback systems, signals sent by a cell to its outside environment, complex intracellular messaging systems and complex membrane receptor/ gateway systems. The messengers are variously proteins, hormones and other biochemicals. Even the simplest free-living cell is a very complicated, precise and highly organised computer-like structure.

23. In simple terms, nervous and hormonal networks interact with body cells and processor node(s). In some cases the central node is called a brain; from this ramifies a network of communication.

24. No. No in principle and practice but necessarily yes, by speculation, according to evolutionary theory. Dormant plants have no brain yet promote growth and respond to a large variety of stimuli. The process is cyclical and homeostatic. Receptors 'engage' with the stimulus and relay their signals by way of chemical messengers (plant hormones and others) to the appropriate cells for response. The response is called a tropism. Whether cellular or multicellular the systems are sophisticated, accurate and efficient. Informative systems are the top-level. Can such a system slowly accrete until all parts are present and it starts to work? No algorithm has been suggested to show how.

25. They turn chemical or light energy into a respiratory substrate called glucose. Both systems are complex. Both need a complementary system call respiration to produce *ATP*. The photosynthetic and respiratory systems are beautifully reciprocal. The former builds

glucose and the latter, using an equally complex suite of codified molecules (including universal *ATPase* that involves a molecular dynamo!), breaks down its substrate to produce *ATP*. Without *ATP* no organism could survive for a minute. It is a 'universal bio-match' that produces a fixed amount of heat to exactly 'ignite' molecular reactions. Never say always and never say never in biology; in this case there are a few modifications found to parts of the system. For example, some microorganisms without access to sunlight use a process called chemosynthesis, as complex as photosynthesis, to obtain *ATP*. But in concept, codification and broad outline he biological driver is the same. Can you suggest how energy metabolism (or any other metabolic pathway with a specific, linked-in end-product) evolved gradually?

26. Darwin thought a cell was simple. He was unaware of molecular biology and information theory. Thus he chose phenotypic species as his measure. Accordingly, he asked 'How does a species first arise and then progressively transform into another?' So the issue becomes the origin of species from which further and (his guess and great mistake) unlimited plasticity in speciation. *The right question is, we now understand, the origin of informative code.* Unfortunately, many still try to squeeze their speculative answers into a Darwinian box. This may be what scientific materialism, the religion of science, wants but not how nature actually works.

27. Yes, continually and ubiquitously. Indeed, nowadays mutation, unknown to Darwin, is judged to be the engine of theoretical evolution.

28. Unlimited plasticity involves gradual, unlimited transformation of one species of organism into another; limited plasticity sets a boundary to this process. In broad terms, unlimited plasticity allows macro-evolutionary process; limited does not.

29. No evidence whatsoever that intensive artificial breeding over the 200 years for changes that improve a breed (except in specific ways humans aim for) have bred improved organisms as a whole or changed one type of organism into another. Minor variation (speciation) has occurred but mainly this involves, due to inbreeding, loss of alleles, that is, of genetic potential for variety by sexual reproduction. Indeed, as Cruft's and other examples show, pushing a types to the limit tends to produce highly strung, sick and

unviable specimens. There is no evidence for change of one type of organism into another; nor, consequently, for the origin of evolutionary innovation. In fact, natural experiments involving millions of generations of *E coli* bacteria, malarial plasmodia, HIV virus, fruit flies and so on only demonstrate limited plasticity. Whatever has been thrown at such organisms in the way of induced mutations and poisons, they remain true to type. This is actual evidence for limited, as opposed to Darwinian speculation for unlimited, plasticity.

30. Palaeontology is a large subject involving many contentious issues. No-one doubts the clear, consistent order of the first, abrupt and fully-formed appearances of major fossil groups and their successions. However, interpretation of the facts differs (as it often does in science) according to the application of materialistic or holistic mind-set. The former allows unlimited plasticity, the latter does not. The former, which treats the unlimited plasticity of macro-evolution as an uncontentious fact, allows argument only within that 'box'. Thus, to fully explore both interpretations and their relative strengths and weaknesses with an open mind it will be necessary to have an expert from each camp go through each mutually contentious bone. This is a task that evolutionary Darwinism, which holds the purse-strings in this area, will probably refuse to endorse. For more information check The Science and Philosophy series, also works from Stephen Meyer ('Darwin's Doubt'), Douglas Dewar (The Transformist Illusion), The Discovery Institute, proponents of The European Model and many others.

31. These are critical for evolutionary theory. Many examples have been hotly contested but none are certain. No-one knows, as in the case of all phyla, how the first specimens appeared. Nor do we know this for even, say, ammonites, but we do know that their well-known fossil succession recycles from simple to complex and back in a way reminiscent of limited plasticity; and that hints of simple first ancestor and eventual evolution into another kind of form are absent. Nor is extinction other than the opposite of evolution. Probably the most hotly contested, ever-changing and dubious interpretations of fossils involve 'human evolution'. However, to lay fossils in an order of similarity is not to prove that one descended from another; and beyond arguments about speciation and phenotypic transformation billions of mutational links are missing,

presumed theoretically to have occurred but unobserved, uncharted, held on trust by faith in theory alone.

32. Reason in reverse. Mindless 'development' depends, by definition, on randomness or, at best, constrained degrees of randomness. In fact, however much a theory may object, reasonable designs - especially complex, specifically targeted systems - are not logically created by chance events. Recognising this problem materialists have, since the time of William Paley, employed the mathematics of probability as clever fingers in a leaking dyke. Possibility, however remote, that something *might have* happened is then slipped a verbal Mickey Flynn converting calculated possibility to an imperative (it *must have* happened somehow because what alternative is there?) Who, of course and after all, can explain with any precision what countless unseen chances did, do or will produce? The matter is incalculably vague. However, such 'reverse-rationality' has by now woozied a dozen generations. It is the ubiquitous but irrational staple of evolutionary explanation for bio-design.

33. None. Except neutral or deleterious impact on forms already present. Randomness innovates nothing logical, meaningful or reasonable. Yet biology is packed with codified reason; and such program is the very antithesis of non-purposive aggregation. How can codified and integrated sets of reasons originate by randomness? The idea is, while indispensably adopted by Darwinian theory, most irrational because its transformations derive, rather than from archetypal bio-logic, from unchartable and thereby unarguable chance. Yet the chancer seems to get away with it. Especially if he lays chance at the feet of a powerful reason, biological survival. In which case he employs the illusory language of animism - the pretence of 'design' by natural selection, mutation or other form of wishless, witless 'guidance'. What, though, *is* survival? Why should aggregates of molecules chance into reason's shapes? Unless, in animism's way, these atomic bodies *'wanted'* to survive. But what in you *wants* to survive? Is the desire psychological or physical? No mechanism or machine, biological or otherwise, *wants* to survive; and unwanting reflexes are codified *for reason of* survival. What machine has ever *wished* itself into its own dynamic operations? No machine, especially no codified machine, occurs at random. Ever. This is fact as scientific as it is non-scientific.

34. A mutation is an accidentally forced change to genetic code. You might expect bugged, garbled code to degrade or, at best, not affect meaning - but not to improve it. However, genetic mutation is the evolutionary G.O.D, that is, a generator of diversity - a radical scientific role-reversal indeed, rightly or wrongly reducing in every sense any 'old-fashioned' concept of the creator of life! Such randomising engine of evolution has to improve the viability of an organism so that it not only survives but, which is progress, dominates competitors. Since the vast majority of mutations are catastrophic, deleterious or at best neutral in effect we seek that minuscule percentage that could be interpreted as beneficial. Although one-offs may briefly affect viral or bacterial circumstance evolutionary *BM*s must aimlessly but successively accumulate, without being lost or deleted, until they produce a complete (and completely unexpected) innovation. Again, the myriad stories, interpretations, guesswork and heavy criticisms that surround this central feature of neo-Darwinian theory cannot be dealt with here. *Let us simply say that there exists no evidence for this and that the probability stacked against the creation of life-forms in this manner is super-astronomical*; at the same time this very slender thread is all that libraries, universities and materialistic secularity in their entirety hang upon.

35. The fact that weaker specimens are, by accident, environmental disturbance or predatory design, eliminated. This is true and the notion, first proposed by creationist Edwin Blyth as a way that created forms were kept healthy, was adopted by Darwin and turned on its head. Now it killed and helped 'create' fresh and evolving forms. In fact, as an agent of survival but not arrival it creates absolutely nothing new, only culls what is already there. In fact, it cannot operate on a feature until it is present, either in full or, arguably, nascent form. It is a mindless editor, a delete button pressed, a negative agent of death.

36. "Machines", he argued, "are irreducible to physics and chemistry." He is absolutely right. In art, engineering or whatever else chaos is proportional to the planlessness of a creator. Automatic, material forces are unaware and, in this sense, chaotic in their creation of inanimate forms. If you equate 'natural' with 'material' the strange paradox arises whereby any planned device (including ones manifest by the million in biological forms) is 'unnatural'!

203

However, the construction of a plan to achieve an end, however trivial that end is, is definitely mindful. And complex systems built to achieve a specific purpose are the epitome of their creator's mindfulness.

37. Sex and reproductive development. The origin of sex anticipates *different yet complementary apparatus* that is able to deliver offspring for *future* survival of the species. As well as major systems biology textbooks describe and holistic authors highlight many thousands of examples of astounding, programmed and specific mechanisms used by organisms. For example, look carefully at your own codified hand. Or, from the rich and lucky dip of evolutionary enigmas pick just a couple of prizes that happen to spring to mind - precisely integrated jumping gears on the legs of a planthopper called *Issus coleoptratus* and a powerful catapult mechanism in the hind legs of leaf (or, generically, flea) beetles such as *Longitarsus anchusae* or *Psylliodes affinis*.

38. Foxed? You are not alone because no-one has a clue. A few weak best-guesses are all that fill the screen.

39. You'll need a detailed textbook for this. But, including molecules and molecular switches as well as complementary organs and their timely development, it runs to many thousands. All pretty much at once. There is neither use nor evidence for a string of fractional, useless but gradually and aimlessly advancing 'sexual shapes'; or tales of what happened while things were in evolutionary gestation. The whole business of reproduction, whether mitotic, meiotic or bacterial, is thoroughly and precisely codified. And code precedes production. Genetic code is the agent of program. Programs have goals - in this case adult, reproductive form. A goal-less series of garbled scripts or chaotic pictures does not.

40. That the human or any other suitable, neutral bio-archetype simply be polarised into two opposite but complementary sexes. Such binary division from singular, metaphysical neutrality would be logical and explain why two sexes (not three or more evolved types) are the immutable case. This view permits conceptual anticipation - because the *clear goal* of reproduction is, and thus in whole process *anticipates*, sexual maturity. From the start the shape of finish-point is known.

204

41. Developed from a single cell biological forms are thoroughly integrated systems. However, science is by inclination analytical and reductionist. It breaks systems into parts. Except that so-called 'scientific holism' treats them as a whole, including their ecology. Such holism reflects dialectical holism but falls short in that it fails to include informative potential as the key component and, especially, to countenance a metaphysical originator of conceptual planning in biology called archetype (or archetypal memory in universal mind). Thus 'materialistic holism' is, although similar in trend, quite different.

42. Everything in biology is codified. This strongly implies, against naturalistic persuasion, that everything was, except for variable adaptations, previously conceived. But never mind that logic, let us turn to application. A brilliant plan might be to reduce the whole body to single cell then grow it up again! What economy of labour. An egg is the nearest to pure potential you will find in a biological body. It is a top-level instructor from which all that follows will develop. Not evolve but develop according to code. You can throw sperm in for variation, the production of 50% of each gender, for excitement and for male company! And for, in most sexed organisms, absolute necessity.

43. The development of three-dimensional form by algorithmic steps - as happens in biological development.

44. It did not evolve. Development shows clear sign of hierarchically developed code that anticipates its end-product, adult form. Which came first? Egg, adult or both, by concept, together? By what conceivable stages could an egg gradually evolve into an adult? How could one immature stage evolve over (un-reproduced!) generations into the next - unless the egg or infant stage was sexually mature? How did blind evolution and its axe, natural selection, know to preserve any fledgling but useless stage? Adults do not appear without prior eggs so the reverse could not have happened. And how, above all, did blind chance rack up the shapes of organs and of bodies that are needed? Unless it is keeping them quiet, evolutionism has no answers.

45. It did not. Again, evolutionism has no satisfactory answers whatsoever to explain innovations required for the dramatic developments of metamorphosis. At each stage a completely

different kind of body with different instincts is transformed, by the logic of algorithmic code, to the next. How did each non-reproductive stage evolve over a vast number of years into the next completely different one? Reproduction is a central feature of survival and therefore of evolutionary theory. But the two do not mix. No competent inventor, however, innovates without conceiving a way to achieve his desired end. A mechanical engineer allows that purpose is critical to his blueprints. Dialectical holism permits such blueprint stored naturally not on paper but in mind. Archetype is a mental form, a program that includes the spatial apprehension of 3-dimensional shape and its coded transmission by way of psychological morphogene to cell and thereby body. Of course, such holistic explanation may be absolutely rational and right but will the creed of scientific materialism, whose basic *mantra* is that 'evolution must have happened', ever change its veto or its mind?

Chapter 6

1. The first compass of community is universal and absolute. As idea is developed into physical expression, as seed develops into body so community reflects a structure of creation whereby all things, psychological and physical, descend from an Absolute Source. In this sense alone is everything connected or, as some aver, 'is one'. This is the Absolute Community of Essence and Existence. And, as all cells in a body are linked to the seed, so all things are linked orderly. In such rational cosmos there exists a hierarchy of connections.

2. This is the community of cosmos relative to you. Since the position of your mind is not coincident with Peak Experience then your social circles radiate in both directions. Internally they consist in memories and relationships with persons and things you think about - the experiences of your mind-world. Externally they include the archetypal, quantum and bulk shells of body and, beyond that, your immediate circumstance, your wider planetary experience and then the knowledge that there is a starry universe. As regards life-forms there is the society of humankind (you family, friends, local community and worldwide population), of animals (from pets, farms and the whole wild world).

3. You or, more precisely, the third eye, seat of mind whence you survey your world. Check diagram in Chapter 2: Cosmological Bearings.

4. Beyond the biotic your ecology includes non-conscious abiotic space, gases (protective shells of atmospheric layers), liquids (mainly water and salt water) and solids (hot and cold, fertile and barren, changing rapidly or not). If all is projected from a Top, Conscious Singularity and then a lower unconscious one (archetypal memory in universal mind), we are all related. Cosmos is, effectively, our 'wider body'. So yes, your perspective should include all of creation. Do not the mystics most pragmatically exhort, " Love without limit"?

5. Ideal Health (*Vis Medicatrix*), dynamic equilibrium and cacophonic pain.

6. We can write:

↓ ill-health   Ideal Health   dynamic health ↑   to which, add:
  pain   Archetype   pain-free body
  blockage   Equilibrium   fluency

7. Liquid water, stable temperature, stable oceanic pH, stable tilt following a practically circular orbit giving seasonal change, stable oxygen level in the air, lunar tides, solar distance from earth and radiation spectrum (70% in visible and another 30 just below it) and other features sum to a dynamic equilibrium with small margin for non-lethal change. This stability has, according to evolutionary theory, obtained unbroken for an extraordinary hundreds of millions of years or else life would have been wiped out. As it stands, earth's biophilic box is ticked.

8. Providers (plants), consumers and recyclers. Provision, consumption, recycled waste. This symbiotic triplex, as James Lovelock well describes, underwrites the balance of life on earth.

9. Bacteria, fungi, worms, some snails and insects return essential nutrients to the ecosystem; our dependency on them is critical. Without them we would (not) be walking on a pile of bodies and detritus miles high out into space.

10. Nature, being non-conscious, is amoral. It is oblivious of feeling. However, when accident, illness or other harsh condition occurs humans sometimes unwarrantedly blame these natural

'catastrophes' for the pain they are suffering. Such pain is 'nature's eyeless negativity'.

11. Cruelty, criminality, evil.

12. One denies a Creator, the other does not. However, atheistic nihilism does not necessarily make for depression and evil - though it certainly helps. There are many atheists, humanists and agnostics of excellent moral character - thoroughly good people. Whether or not the principles that mark this goodness were transferred from long Christian tradition is a point for interested academics to debate because, in fact, it is the 'now' of behaviour that counts. In this case, the difference in character between honest, well-meaning theist and atheist seems as paper-thin as a metaphysical 'credo'; but there is a gulf in perspective regarding the true nature of creation.

13. Religion, politics and law.

14. ↓ incorrect alignment   Informative Principle
    correct alignment ↑

    also:

law	Religion	politics
control of disorder	Ideal Government	orderly business
principled enforcement	Justice/Balance	principled legislation
immorality	Moral Ideals	morality-in-practice

15. Enlightenment explained to the unenlightened. Its ecstasies can be experienced but otherwise the only way to describe what is metaphysical is by metaphor, say, poetry or parable. And the tendency of the mass of us is, in any venture, to establish, as far as a group's understanding goes, habit, rhythm and stability of practice. Some even convert explanatory story into fact. Thus creed and ritual gradually encrust and formalise the intense enthusiasm at religion's core; but, led more or less successfully by different hands, they form a vehicle in which High but Hidden Truth is borne. A religion's Essential Nobility is enduring but its structures will, in the nature of existence, wither or die. Even occasional, local perversions, oppressive or painful but short-lived, can erupt; but the core heat also regularly breaks out, refreshes and reforms.

16. Politics is the ego. It is centaur man, who looks both ways, at work. On the one-hand, it takes principle from its country's ideals and translates them into legal practice whose application guides and affects a country's social life. On the other, economics of the body dominate; survival, trade and material quality of life force to the

fore. Business in this department tends towards vice; lust, wealth, power, passion and self-interest work here. Only a 'sense metaphysical' restrains - rules of morality that curb the buccaneer. History's panorama illustrates how, principled, ill-principled or unprincipled, a government that's crewed by centaur men behaves. Their tendencies can drive a country on the rocks or, ideally, navigate life's social seas successfully.

17. Because its duties run counter to lesser self, *ego* and physical desires. The Self is not self and thus, as self is gradually dissolved and Self approached, love and compassion automatically, like cream on milk, arise. On this basis Absolute Morality would be as easy as Love. In fact, it would *be* complete love and compassion.

18. Illumination. If the nucleus is Self, Psyche or Soul then psychology and psychiatry are, except in debased or intellectualised forms, Ways of Light. The goal is to discover Knowledge, Wisdom and Central Truth. Natural Dialectic lays out orderly construction of the cosmos and, therefore, the metaphysical path to this Heart.

19. Physical associations are everywhere about. There is, however, contemplatives aver, inner metaphysical association. Yes, there are images, imaginations, dreams and emotions peopling every individual's mind. The quality of these associations affects mood, decision, purpose, plan and maybe physical activity. Yet it is also possible to contemplate in search of understanding and perhaps answers to the universal questions some of which we met in chapter one. Still further up the Mount Universe resides its Peak - the conscious source of the entire cosmos found below. And what is conscious is alive; you can relate to, love and merge in what is live. So 'individual association' leads towards connection and communion with this Centre. Know Thyself. Millions of men and women strive to reach this Grail, Stone, perennial Elixir. Obscured by cloudy and perpetual machinations of the mind they practice to reach nature's Nature - their own Top Centre which is at the same time Central Light Source of the universe. To disperse the mists of mind and reach this Infinite yet Single Centre is the mystic path.

20. By nature immeasurable, inexplicable and ineffable. Who has experienced Supreme Being? However, for aspects/ qualities that point towards understanding read, in any stack, the right-hand, capitalised elements of Primary Dialectic.

www.ingramcontent.com/pod-product-compliance
Lightning Source LLC
Chambersburg PA
CBHW052145070326
40689CB00050B/2010